AGRARIAN QUESTIONS

Lawrence Busch, General Editor

SCIENTISTS
IN THE
THIRD
WORLD

Jacques Gaillard

THE UNIVERSITY PRESS OF KENTUCKY

This book is a revised and expanded edition of *Les Chercheurs des Pays en Développement*. Paris: Institut Français de Recherche Scientifique pour le Développement en Coopération (ORSTOM). 220 pages.

Scholarly publisher for the Commonwealth, serving Bellarmine College, Berea College, Centre College of Kentucky, Eastern Kentucky University, The Filson Club, Georgetown College, Kentucky Historical Society, Kentucky State University, Morehead State University, Murray State University, Northern Kentucky University, Transylvania University, University of Kentucky, University of Louisville, and Western Kentucky University

Editorial and Sales Offices: Lexington, Kentucky 40508-4008

Library of Congress Cataloging-in-Publication Data

Gaillard, Jacques, 1951-
[Chercheurs des pays en développement. English]
Scientists in the Third World / Jacques Gaillard.
  p. cm. — (Agrarian questions)
Rev. translation of: Les chercheurs des pays en développement.
Includes bibliographical references and index.
ISBN 0-8131-1731-3 :
1. Scientists—Developing countries. 2. Research—Developing countries. I. Title. II. Series.
Q127.2.G35   1991
509.172'4—dc20   90-47574

This book is printed on acid-free paper meeting the requirements of the American National Standard for Permanence of Paper for Printed Library Materials. ∞

Contents

Tables and Figures

Figures

Acknowledgments

The aim of this book is to contribute to the reader's understanding of how, by whom, and under which conditions research is actually conducted in the developing countries (DCs). Although the main focus is on DC scientists who received research subsidies from the International Foundation for Science (IFS), the results will be useful not only to IFS and other research aid and scientific development organizations, but also to researchers and science policymakers in general.

Although the major part of this study was written between 1986 and 1988, the background work was started toward the end of 1984 when I was still a scientific secretary at IFS in Stockholm. The original inspiration dates back to discussions I had with my friends Geoff Oldham and Fransisco Sagasti one evening in 1982. Afterward, at a seminar held at the Research Policy Institute of Lund in September 1984, I started discussing an early draft with the late Michael Moravcsik, and then, a bit later, with members of the ORSTOM (Institut Français de Recherche Scientifique pour le Développement en Coopération) research team on science, technology, and development in Paris.

Several people including Rigas Arvanitis, Jacques de Bandt, Lawrence Busch, Yves Goudineau, Said Ouattar, Michael Pollak, Jean-Jacques Salomon, Albert Sasson, Pierre Tripier, Roland Waast, and, most important, my wife, Anne Marie, read and suggested numerous improvements to a draft version. Hervé Chevillotte, former head of the ORSTOM Information Service at Bondy, gave me invaluable help in processing statistical data. Francine Sinègre, the ORSTOM librarian at Bondy, was very efficient and helpful in digging out bibliographical references. Tilly Gaillard did yeoman's work in translating it. The intellectual and moral support I received from Sven Brohult, founding president of IFS, was decisive in bringing the study to fruition. The backbone of this book comes from the IFS grantees themselves. Without their answers to my questionnaire, and the many very enlightening discussions we had, this book could not have been written. My sincere gratitude goes to all these people.

I would like to express special appreciation to the Salén Foundation, ORSTOM, and IFS for their financial backing.

Preface

Developing Science and Training Scientists
in Third World Countries: A Must

The scientist's golden age of generous patrons is gone. Science is becoming more and more costly. So costly, in fact, that some of the smaller industrialized countries can no longer fund research teams, build and equip laboratories, and then keep them running. This being the picture, how can developing countries (DCs) justify investing in research? They are confronted with more pressing priorities, and in the past, research and science have not generated the expected development.

The official discourse on "science for development," heard especially from the United Nations Educational, Scientific and Cultural Organization (UNESCO) in the 1960s, insisted on the need to increase resources for scientific research in the DCs. An international conference held in 1964 in Lagos led to the "Lagos Plan," which contained a series of recommendations designed to further scientific growth and independence in the DCs. The two main recommendations stated that (1) each country should immediately devote 0.5% of its Gross National Product (GNP) to research and development (R&D), and the figure should be increased to 1% by 1980; and (2) each country should aim to train 200 scientists for every million inhabitants by 1980.

In many countries, the latter objective has been reached or even exceeded, especially in Asia and in Latin America; but, except for a few countries in Asia, we cannot say the same for the former (Gaillard 1986). The principle of devoting greater resources to research was reasserted at the United Nations Conference on Science and Technology for Development held in 1979 in Vienna. Emphasis was placed on the need to develop "endogenous" scientific and technological capacities in harmony with the "social and cultural traditions" and the "particular conditions of each DC." The DCs were well aware of the need to enhance their scientific and technological potential. Most of them established departments for scientific coordination and decision making at the highest level of government, inspired as they were, with a bare 10 to 15 year lag, by the dominant models of the Western countries.

Perrin (1983, p. 61) wrote, "Buying technology does not mean controlling it." Similarly, "science does not guarantee development" (Salomon 1984). In other words, it is not enough to invest in R&D, to set up research teams, and to build universities and laboratories with libraries and sophisticated equipment for miraculous scientific discoveries to happen and development to bloom. A scientific policy is only a component of a logical whole, one of the pawns in a national development strategy. This being the case, isn't it obvious that the DCs do not have the conditions required for autonomous development? Isn't science an intellectual consumer good available to an elite group, something that will never be able to contribute to national economic and social development? Isn't it easier and cheaper to import the scientific knowledge needed for development and thus save the price of a national research system? Isn't it better to recognize that from now on research is a luxury, within the means of only the richest Western countries?

We think that this would be a wrong choice and that each country should feel that building up an endogenous scientific community must be a priority goal. Western science should not be considered as an exclusive model, and each country should adapt its research system to its socioeconomic conditions and development strategies. It also seems important to emphasize that developing science is a lengthy undertaking. Even in much more favorable conditions, countries like the United States and Japan needed more than 50 years to develop a scientific potential that could stand up to Europe. Western science started making inroads in Japan in 1869 (Price 1963), but it was not until the 1960s that Japan's scientific power had emerged to a level that attracted world attention.

On this subject, Moravcsik and Gibson (1979, p. 28) were right to stress that as time goes by there is a serious decrease in the predicted scientific potential, especially since many countries must start from scratch and must create everything: "Of young students selected for education to become a scientist, many will never attain the advanced degree; of those who do, many will never produce anything beyond their thesis; of those who do, many will produce only a few additional pieces of work, after which they 'die scientifically'" It takes about 20 years of training before a scientist can become productive. This is a long time. A historical perspective composed merely of the last 20 to 25 years is not long enough to draw reliable conclusions on the emergence or nonemergence of endogenous scientific communities, of the right size, in the DCs. The UNESCO Statistical Yearbook (1985a, V–9) reports that the increase in scientists and engineers and the increase in R&D

Table P.1. Distribution of Scientist and Engineers, and Funds for R&D, as Percentages of World Totals in 1970, 1975, and 1980

Year	Scientists & Engineers (%)		R & D Expenditure (%)	
	Developed Countries	Developing Countries	Developed Countries	Developing Countries
1970	92.1	7.9	97.7	2.3
1975	91.1	8.9	96.1	3.9
1980	89.4	10.6	94.0	6.0

expenditure in the DCs is relatively high compared to the rest of the world, for the period between 1970 and 1980 (table P.1).

This table first and foremost shows the uneven distribution of R&D resources in the world and their vast concentration in the industrialized countries. Again referring to the UNESCO mean figures (1985, V–13), we see that there are close to 24 times more scientists and engineers per million inhabitants in the developed countries (2,875) than in the developing countries (121). Similarly, R&D outlay as a percentage of the gross national product (GNP) would be equal to 2.24% for developed countries as a whole and 0.43% for the DCs. These figures, moreover, hide even greater disparities that are drowned in global statistics. A more thorough analysis of the situation in each continent and each country is relatively more favorable to Asia and certain countries of Latin America than to Africa, which is at the bottom of the list both in scientific manpower and in R&D funding. But definitions change, and the borderline between developed and developing countries is not always easy to define. The staging of new concepts such as Newly Industrialized Countries (NIC) is a good example, and we might start thinking about creating more categories like "underdeveloping countries," a term for countries that are dropping out of the category of developed countries. We must also be careful not to apply a simple rule of proportionality when comparing funds allocated to research, because the "scale" factor certainly must place a handicap on the small countries in any comparisons made with the capacities of the major scientific powers.

The last statement no doubt stokes the fires of those who claim that the idea of creating a national research capacity in the DCs is nothing more than an illusion. Actually, it is true that because of their size and available resources, very few DCs will be able to contribute to the production of scientific knowledge and, in particular, to new technologies. This inspired Salomon (1986) to make a distinction between controlling production and controlling the use of scientific knowledge and technological change. Even people who do not see the need for each country to have its own scientific community recognize the vital importance for each country to have access to technology. But the question is whether technology can be successfully transferred without a minimum of scientific expertise at the receiving end. The answer, obviously, is no. Furthermore, much of the knowledge and technology conceived in the developed countries cannot be applied directly in the DCs. There are always problems connected to knowledge appropriation and technology transfer that can only be solved through on-site science—in other words, on-site scientists.

Another justification that seems of prime importance is related to the fact that all modern nations need to have their own higher education systems to train their senior employees and that it is quite impossible to conceive of higher education without teachers who are actively involved in research. This is the only way to keep them in contact with science on-the-move and thus keep them up to date on recent progress in their respective disciplines. All too many DC university professors no longer practice science and merely recite scientific facts of the past instead of teaching their students that science is a method for stating and solving problems. This same attitude is still too often heard even in the industrialized countries. We could find a whole series of arguments to justify promoting science and technology in the DCs, but that is not the subject at hand. We feel that we are going far enough by concluding simply that the nations that staunchly chose to embark on experimental science now number among the richest, and the DCs have every right to seek their share; they do not want to be left in the wings of the scientific revolution that is now being acted on the stage before them.

Scope of the Study

Now that the stage has been set, there are still some problems of direct significance to the role of science in the development of the DCs that we will not be able to deal with thoroughly in this study. My mind turns especially to the debate on priorities, which takes us straight to the matter of training for scientific and technological personnel and to strategies of development. What science is needed for what development? Answers to this question are both emotional and contradictory. Is the future of the DCs embedded in the new technologies that have mobilized only a very small number of them, such as Brazil and the Republic of Korea, or is it in the agricultural R&D that involves more than three-fourths of the populations of Africa and Asia? The two approaches are not mutually exclusive; new technologies are not incompatible with agriculture. Should we only encourage applied research in a finalized form that can answer urgent needs and provide fast answers to the problems of underdevelopment, and consider basic research a useless luxury?

Because of the growing differences between DCs, strategies cannot be standardized; there is no universally applicable approach. We are convinced that recourse to science, and thus to scientists, must form part of these strategies.

This study talks about these scientists and the world in which they work.

Chapter One

Introduction

Studies on Science and Scientists in Developing Countries: Abundant Documentation but Fragmentary Knowledge

There is a relatively large storehouse of documents and reports on science and technology policies in the developing countries that have been prepared, more often than not, for international conferences such as the 1979 United Nations Conference on Science and Technology for Development held in Vienna (1980). This said, we have to recognize that official speeches mainly contain statements of intent and that our knowledge of science and scientists in the DCs is very incomplete. There is a relative abundance of literature on DC science scattered through numerous journals, seminar reports, and proceedings; but there are far too few empirical studies. Moravcsik (1976) has written one of the most complete bibliographies on this subject, covering literature up to the early 1970s. We share his feeling that research on DC science is an unexplored and fruitful area (Moravcsik 1985a).

Generally referring to national statistics compiled by organizations such as UNESCO and OECD (Organization for Economic Co-operation and Development), certain authors have emphasized the shortcomings of the DC research systems and the shortage of available resources (Rossi 1973). Other authors compared socioeconomic conditions with the level of scientific development in these countries (Eres 1982). In some of his writings, de Solla Price (de Solla Price and Gursay 1975) gives quantitative indicators for the DCs. Research by Garfield (1983) and his Institute for Scientific Information (ISI) in Philadelphia points to the low productivity of DC scientists, the difference in productivity levels (half the work published by all DC scientists comes from India), and degrees of dependency (articles by DC scientists have greater impact when coauthored with scientists from developed countries). Using the ISI data base, quantitative analyses of mainstream scientific literature, i.e., articles in internationally read publications, were made at the continentwide and national levels. One such bibliometric study on mainstream science in Singapore recently showed that articles written by national scientists and published in international journals were very rarely cited (Arunachalam and Garg 1985). Care must be taken not to overlook the value of this bibliometric approach that relies on the choices reflected in the data base. This method should only be used to examine international mainstream literature, since, when it was used on the DCs, it gave the impression that approximately 5% of the world's scientific output was to be credited to these countries, while bibliometric work carried out at ORSTOM on soil sciences and agriculture suggests that over 20% of the scientific production concerning these disciplines in "the hot regions of the globe" comes from the DCs (Chatelin and Arvanitis 1988a). Other authors such as Frame et al. (1977) provide interesting general information on the respective ranking of various DCs and on the distribution and orientation of scientific disciplines.

There has been little research on the scientists who make up the DC scientific communities or on how these communities are developing. The comparative study by T.O. Eisemon (1979) is among the exceptions. He interviewed teachers/scientists in the mathematics and zoology departments at the universities of Ibadan and Nairobi in 1978 and concluded that "the achievements of Nigerian and Kenyan science are primarily quantitative and in the sphere of construction of an institutional framework for scientific research. Science teaching programs have been developed, scientific societies established, publishing institutions formed. These are not trivial accomplishments in my view. Nevertheless, it is also true that scientific work—in a more substantial sense—has not been much advanced. . . . Nor have hopes for rapid scientific development been realized. A much longer time will be required before a conclusive judgment can be passed on the effective implementation of the scientific 'ethos' in Black Africa." Another study on science in Mexico suggests that, as in the other DCs, there are fewer scientists than purported. The study concludes by saying that the "Mexican scientific community is like an army which has too many generals and too much equipment but which lacks soldiers, particularly well trained soldiers" (Schoijet 1979, p. 404).

In reality, most of these studies tend to show that none of the DCs has a genuine scientific community, not even India, which numerically has the world's third largest scientific community (Shiva and Bandyopadhyay 1980), or Brazil (Schwartzman 1978). What an Indian physicist had to say about research practices in his country was most revealing: "There is no scientific community in this country. . . . I meet my colleagues only abroad. I meet my colleagues even from Delhi abroad. . . . In a well-knit community, where you are exchanging preprints, things are happening and there is excitement. There is no excitement here. Our excitement comes by mail from outside. It depends on the postal system. This is the worst part; the spirit is dead" (Shiva and Bandyopadhyay 1980, p. 587). This dependence on the outside environment, in other words the West, is an often sung theme song for many DC scientists. Consequently, since the knowledge formation process in the DCs is largely influenced and determined by the Western world, a considerable part of their scientific output is foreign to where it is produced.

Problem Orientation

These recent studies confirm what Stevan Dedijer (1963) felt in the early 1960s when he wrote:

In the underdeveloped countries scientists are relatively few in number, and they are often, as far as any particular field of research is concerned, dispersed over long distances. They suffer from isolation from each other, and thus they do not have the benefits of the stimulation of the presence of persons working in closely related fields. They are in danger, a danger to which they too often succumb, of losing contacts with their colleagues in the international scientific community. They feel peripheral and out of touch with the important developments in science unless they can visit and be visited by important scientists from the more developed countries; they feel inferior and neglected because their own journals and organs of publication, where they exist at all, are seldom read by foreign scientists, seldom quoted in the literature and are indeed often neglected by their own colleagues at home. They have little contact with their colleagues in neighbouring underdeveloped countries. They are in brief not fully-fledged members of the scientific community and their work suffers accordingly. [80–81]

More than twenty years have gone by since these words were written, but we have to admit that they somehow still ring true.

It is now generally recognized that an endogenous scientific community can only develop in a peripheral position if its members have sustained relations with the center. One of the best ways to establish such relationships is through interpersonal contacts between scientists who come together because of their common interests and mutual respect (Herzog 1983). The ways to establish such relations include studying at foreign universities, often in a country in the center; attending conferences; taking sabbatical leaves to conduct research abroad; and corresponding with foreign scientists and publishing articles in journals of international repute.

How often do DC scientists use these mechanisms? How effective are they, and what are their effects on science as it is practiced in the

DCs? What are the related problems? Is it necessary for DC science and scientists to rely on the center for training, for their choice of research subjects, for funding research; and must DC scientists belong to the international scientific community. Are these preconditions to creating national scientific communities in the DCs? Are there other ways?

Our experience in practicing research in the DCs leads us to believe that the most productive scientists with the most promising results are the ones who strive to form active groups of researchers focusing together on a high-impetus program and who participate in setting up effective scientific institutions. These are the people who have had the opportunity to do most of their studies abroad and have established and maintained the strongest personal contacts with experienced scientists from the most advanced Western countries. As a result of their many years abroad, they are most affected by Western research methods, subjects, and models, even if these latter are not always applicable in their home country. These scientists are also the ones who have lost contact with the everyday realities and problems of their countries and who often feel closer to the Western scientists and values than to the people and traditional values of their home society. And as they are whirled up the career path, they find themselves all too rapidly in high-ranking posts, consumed by administrative and bureaucratic duties that prevent them from practicing science at a time in their life when they should be at the height of their productive career. And finally, these scientists, as a result of their continued contacts with foreign sources, their qualifications, and their difficulty in readapting upon return home, will be solicited most often to accept better paid employment abroad in working conditions that they grew accustomed to during their stay abroad.

We intend to provide partial answers to this situation and confirm or refute the above through reference to a study carried out on 489 research scientists working in 67 DCs. Our report will also discuss the sociocultural origin of the respondents and the formation of a DC intellectual class. Attention will also dwell on their job motivations and their perception of their work and research methods in their home country in general.

These scientists have in common that they all received at least one grant from the International Foundation for Science (IFS) some time between 1974 and 1984. As an IFS scientific secretary, I had the opportunity to meet most of these grantees during my travels to some 60 DCs or at seminars and conferences where they reported on their work. The following brief description of IFS activities will facilitate understanding the specific characteristics of the survey population and the system of recruitment.

The International Foundation for Science

International organizations and private foundations in the OECD and the COMECON (Council for Mutual Economic Assistance) countries have developed numerous programs to train scientists from the DCs (Gaillard 1979, 1984a, 1985a), but most programs nearly always offer the first years of study in specialized universities and organizations in the industrialized countries (ICs).

What is unique about IFS is that it helps young graduates from DCs work within their home institutions to set up research programs in biological and agricultural sciences applied to the basic necessities of life, and technology applied to rural problems.

IFS was created in the early 1970s, and by the end of 1988 close to 1,500 young scientists in over 80 countries of Asia, Africa, Latin America, and Oceania had benefited from its grants.

The idea that inspired the creation of IFS was developed in the 1960s and was discussed by the Pugwash group (named for the Canadian city where the group first met) in 1965 and in 1969. The initial discussions brought together eminent specialists such as Robert Marshak, nuclear physicist at the University of Rochester; Roger Revelle, director of the Harvard Center for Demographic Studies; Abdus Salam, director of the International Centre for Theoretical Physics in Trieste; and Paul Auger from UNESCO. The chief proponent was Sven Brohult, president of the Swedish Academy for Technical Sciences (IVA). Thanks to a subsidy from UNESCO, two Swedish academies were able to organize a three-day meeting in July 1970 attended by 32 representatives from scientific academies and research organizations from 10 ICs and 6 DCs to discuss problems related to the development of science and the working conditions of scientists in the DCs. At this meeting in Stockholm, a committee, presided over by Professor Paul Auger, was set up to prepare the creation of an organization to be called the International Foundation for Science (IFS).

During ensuing discussions, the members of the committee expressed very serious differences of opinion concerning the size, structure, and functioning of the organization. There were two groups representing two lines of thought. One wanted IFS to be connected to UNESCO. The other, composed primarily of academicians, wanted IFS to be an independent, nongovernmental organization; and thus it was

decided. In January 1972 a letter was sent to some 50 research academies, institutions, and the like, inviting them to become founding
members of IFS. On March 25, 1972, with 14 founding members, the
interim committee announced the official establishment of IFS. An
interim Board of Trustees was established with Prof. Sven Brohult as
chairman. Contributions from the Salén Foundation (a private Swedish
business foundation) and the World Bank made it possible to set up a
small secretariat. Initial financing, in 1973 by Sweden and Canada, was
used in 1974 to pay for the first 45 research grants.

By the end of 1988, IFS had 85 member organizations in 71 countries.
The aim and objectives of IFS have not changed since 1972 when they
were first defined in the IFS statutes: "promote and support high
quality scientific and technological research in the DCs in the exact,
natural and social sciences selected because of their importance to
national development . . . support worthy young DC scientists and
technicians by providing financial and other types of support. Grantees are selected on the basis of the quality of their projects, the value of
these projects for the country and the region. . . . The research must
take place in a DC. . . . The grants must be applied for by individual
scientists."

In 1981 the IFS Sponsors Committee charged an independent panel
of experts to evaluate IFS activities (Sagasti et al. 1983). These experts
emphasized that IFS was unique in that it was the only institution
whose sole aim was to offer direct support to individual scientists at the
beginning of their career. The maximum for an IFS grant was set at U.S.
$10,000. One of the prerequisites for obtaining a grant was that the
applicant had to be employed by a DC university or research institute
that agreed, after the grant had been authorized, to provide the grantee
with a laboratory or other basic facilities required to implement his
project. Grants could be renewed three times. Applications for renewals were judged on the basis of the quality of the grantee's work
during the preceding research period. It has happened that IFS grantees, although usually young graduates, find employment as established research scientists or are rapidly promoted to senior positions
such as dean of a faculty, vice-chancellor of a university, or director of
an institute, or are even put in charge of coordinating research at the
national level, after only one or two IFS grants. When this happens, IFS
no longer provides support.

By helping young DC scientists mature, IFS is filling a large gap in
the network of international organizations working to support scientific and technical research activities in the DCs. IFS is singular in its

Table 1.1. Types of S&T Research in DCs and Sources of External Funding

Types of S&T Research	Requirements and Characteristics	Range of Funding	Funding Agencies
Individual research efforts	scientific capabilities and creativity interaction with peers access to literature and travel limited access to scientific equipment and materials one scientist with (possibly) technical assistants minimum of managerial skills 1–4 years	under US$ 10,000	IFS Private foundations, e.g., Ford, Rockefeller
Research projects	scientific and technical capabilities and creativity interaction with peers and users of research results access to literature and travel steady access to specilized scientific equipment and materials multidisciplinary teams (e.g. 3–5 researchers plus assistants) access to (possibly) pilot plants and small-scale field trials intermediate level of managerial skills 2–5 years	US$ 10,000-150,000	SAREC BOSTID UNESCO UN Agencies UNDP NUFFIC IDRC CIDA SIDA ORSTOM GTZ Other bilateral agencies (Dutch, Italian, Belgian, etc.)
Large-scale research and implementation programs and projects (possibly including extension work, industrial implementation, and training)	technical and scientific capabilities interaction with other research groups, farmers, industry, government agencies, international agencies, etc. access to literature, travel and frequent field trips access to specialized laboratories and materials exclusively for the program or project large multidisciplinary teams (e.g. 5 to 20 researchers plus assistants) access to pilot plants and industrial facilities, large-scale field trails high level of managerial skills and control 4 or more years	over US$ 150,000 (in some cases several million US$)	WHO CGIAR World Bank IFAD ADB IDB

Source: Sagasti et al. (1983), Table 1, p. 7.

system of supporting individual scientists (see table 1.1); other organizations finance research and development programs. This division of roles is especially well accepted since the other organizations referred to above number among the main sponsors of IFS, viz., the Swedish Agency for Research Cooperation with Developing Countries (SAREC), Canada's International Development Research Centre (IDRC), and the United States Agency for International Development (U.S.AID).

During the last few years, the IFS budget has stayed at about U.S. $2 million. In 1984 contributions were made, in descending order, by Sweden, the United States, Canada, the Federal Republic of Germany, France, Australia, the Netherlands, Nigeria, Belgium, and Switzerland.

The typical grantee is between 28 and 40 years of age. About 60% have a Ph.D. and 25% an M.S. Approximately 16% are women. The average IFS grant has been steadily increasing since 1974 and in 1984 amounted to about U.S. $7,500. Grants are intended to cover the cost of equipment, supplies, documentation, travel, and, if necessary, technical assistance and labor. IFS will not pay the grantee's salary and will not support secret or military research. In over half the cases, the grantees have received either higher amounts than the 1984 average or at least one grant renewal. The home institutions often provide the grantee with resources well beyond what IFS gives.

An applicant must be a citizen of and carry out research in a DC. During the first few years, the majority of the grantees came from Southeast Asia and India. After great effort, IFS has gradually achieved a better balance in favor of Africa and, to a lesser extent, Latin America. Since the beginning of the 1980s, grantee applications have gone up considerably. The IFS Secretariat received more than 400 applications in 1984 and 1985, which means that grants are awarded to about one out of five applicants.

All applications are appraised by a panel of scientific advisers who send in their evaluations by post or express them at semiannual meetings. The conclusions of the scientific advisers are submitted twice a year to the Executive Committee, which takes decisions on the basis of funds available, geographical distribution, the state of research in the country concerned, and the number of grants to be distributed.

Another important part of IFS activities concerns supplementary assistance for grantees. Additional sums may be granted for contingencies, e.g., to cover sudden price increases for scientific equipment, to purchase supplemental equipment, or to pay for attendance at scientific meetings. The IFS Secretariat can also help grantees in contacting suppliers in order to improve the cost/quality ratio, settle invoices, obtain catalogs of scientific equipment, or obtain scientific publications or references not available in the grantee's home country.

Organizing workshops and conferences has become an important part of IFS activities since 1978, when it held its first four. Since that time more than 30 seminars have been held in the four continents where IFS is working, bringing together hundreds of grantees.

In the beginning, discussions on how to formulate the grant program and what research areas should be adopted were very impassioned. Suggestions ran from theoretical physics to biology to medicine and their applications. After study visits and consultations with numerous scientists and organizations in both the DCs and the ICs, the interim committee, in 1974, set out six priority research fields connected to biological and agricultural sciences. The six original, and present, fields of priority are aquaculture, animal production, food crops, afforestation and mycorrhiza, fermentation and applied microbiology, and natural products. In 1978 rural technology became the seventh priority area. These fields were chosen with the expectation that results could have a positive effect on socioeconomic development and improve the well-being of the resident populations.

The Methodological Approach

Our study, which is part of a study on the impact of IFS support on the work and career paths of the grantees, involves four complementary actions.

1. A questionnaire survey sent to the 766 scientists in 78 countries who had received IFS grants between 1974 and 1984.
2. Interviews and discussions with DC scientists.
3. A quantitative and qualitative bibliometric study of work published by close to 200 scientists in the group.
4. A comparative study of three DC scientific communities representing the three main continents and three different historical and political backgrounds, viz., Senegal, Thailand, and Costa Rica.

This book presents the results of the questionnaire survey, some of the interviews, and the bibliometric and comparative studies. It is also largely supported by personal observations and substantial documentation collected during field trips.

The first version of the questionnaire was drawn up in September 1984. It was revised subsequent to discussions with the ORSTOM research team on scientific policies and practices and comments received from Prof. Lawrence Busch of the University of Kentucky. It was tested in October 1984 on a small sample of IFS grantees during the Fourth IFS General Assembly in Rabat, Morocco. The final version (Appendix A) was sent, for the first time, in March 1985 to all the IFS grantees. The questionnaire was drafted in English and then translated into French and Spanish. The Portuguese-language grantees, viz., seven Brazilians, received the Spanish version. Reminder copies were sent out in May and October 1985. We coded the questionnaire in December 1985 and January 1986 and commissioned a private company to computerize the data in February 1986. Data processing and statistical analysis were started in March 1986 at the ORSTOM computer services in Bondy, France, with assistance from its director, Hervé Chevillotte, and from the CIRCE (interregional center for electronic computation), which is part of the French national center for scientific research (CNRS).

Responses to the Questionnaire

Out of an initial population of 766 scientists in 78 countries, 489 in 67 countries, or 63.84%, answered (see Appendix B). The 11 countries from which no answers were received have fewer than four grantees each. The countries were Benin, Botswana, Liberia, Rwanda, Swaziland, Bolivia, Trinidad, Afghanistan, Turkey, Western Samoa, and Tonga.

All the respondents are DC scientists, and all are working in their home country except for eight who are working in another DC; 71.4% work for university or other academic institutions, and 26.2% for national research and development institutions. Coincident with the IFS population as a whole, 83.4% are men, and 16.6% are women; 80% are between 30 and 45 years of age. More than 60% of the respondents have a Ph.D. or the equivalent, for which most of them (76%) studied in a developed country, mainly in the United States (26%), Great Britain (20%), or France (15%). The other respondents have either an M.S. or the equivalent (25%), a "license" or a degree from an engineering or veterinary medicine school. As concerns time spent abroad to obtain these degrees, 50% of the respondents spent between one and five years abroad, 25% spent more than five years abroad, 25% spent less than one year (or no time at all) abroad.

The geographical distribution of survey respondents per continent corresponds almost perfectly with the distribution of the total population, as table 1.2 indicates. The small percentage of Latin American

Table 1.2. Responses to Questionnaire by Geographical Area

	Responses		Totals	
Africa	182	37.2%	291	38.0%
Latin America	86	17.6%	135	17.6%
Asia	213	43.6%	324	42.3%
Pacific	8	1.6%	16	2.1%
Total	489	100.0%	766	100.0%

Table 1.3. Responses to Questionnaire by Scientific Field

Scientific Field	Responses		Totals	
Aquaculture	98	20.0%	135	17.6%
Animal production	75	15.3%	138	18.0%
Food crops	113	23.1%	180	23.5%
Afforestation & mycorrhiza	39	8.0%	56	7.3%
Fermentation & applied microbiology	53	10.9%	85	11.1%
Natural products	90	18.4%	137	17.9%
Rural technology	21	4.3%	35	4.6%
Total	489	100.0%	766	100.0%

grantees can partly be explained by the fact that IFS deliberately avoided publicizing its program in countries that were well on their scientific way, such as Argentina, Brazil, Venezuela, and Mexico, although it is an IFS principle not to exclude any DC from the list of potentially beneficiary countries. Further, up to now IFS has had only two official working languages, French and English, which is a handicap for the Spanish speakers.

The geographical distribution of the grantees reflects a deliberate policy that IFS adopted to extend assistance to scientists working in countries and institutions that had special difficulty in carrying out research, a policy that favored Africa. This geographical distribution obviously does not coincide with the worldwide distribution of scientists and engineers working on research and development in DCs, which shows Asia to have 84%, Latin America 11%, and Africa 4%. These percentages are calculated using 1980 data from the 1985 UNESCO Statistical Yearbook.

In comparison with the initial population, we can see in table 1.3 that the number of responses relative to the various scientific areas are satisfactory. A more detailed analysis would show an uneven geographical distribution of IFS grantees for certain priority areas. In the field of aquaculture, more than half the grantees are working in Asia. This concentration can be explained by the larger numbers of potential scientists and by the existence of a tradition that in certain Asian countries is more than 2,000 years old. Furthermore, over 80% of the world aquacultural production comes from the Indo-Pacific region.

Over 50% of the grantees working in food crops are located in Africa. Without seeking to justify this distribution, we should note that a special effort is needed to reverse the trend reflected in the per capita food production index since the beginning of the 1960s, a trend that grew worse in the 1970s when agricultural production, on a per capita basis, dropped by 1% per annum. For the African continent as a whole, the U.N. Food and Agriculture Organization (FAO) and the World Bank estimate that the 1.5% average annual primary food production increase (the figure was 2% in the preceding decade) does not even keep abreast of population growth. During the last few years, this decline has been continuing, and the production gap between sub-Saharan Africa and the other regions of the developing world has been growing (Gaillard 1984b, p. 33).

More than half the grantees working on fermentation and applied microbiology are based in Southeast Asia. Here again, there is a long tradition. Sauces and foods are often fermented in these countries by skillfully using microorganisms to increase the nutritive value and improve the preservation. The grantees in the other four fields are more or less equally spread out across the three main continents.

The only systematic—and expected—bias in our comparison between those who responded and those who did not was related to the year of first grant, the number of renewals, and whether IFS still supported the research. Thus, grantees currently receiving IFS support tended to respond more readily than those who were not and those who would probably not receive more support in the future, as table 1.4 indicates. The more grants a scientist had received, the more he tended to show gratitude through his relatively greater readiness to answer questionnaires, as table 1.5 bears out. It is also likely that the researchers who received three grants or more could be counted among the most active and productive scientists.

Finally, the more recent the award of the first grant, the more readily the grantees tended to respond. Thus, whereas less than half of the researchers who had obtained their first research grant from IFS dur-

Table 1.4. Responses to Questionnaire by Status of IFS Support:
Active or Terminated

IFS Support	Responses		No Responses		Total
Active	398	66.7%	199	33.3%	597
Terminated	91	53.8%	78	46.2%	169
Total	489	63.8%	277	36.2%	766

Table 1.5. Responses to Questionnaire by Number of Grants
Obtained

No. of Grants	Responses		No Responses		Total
1	249	58.6%	176	41.4%	425
2	143	66.2%	73	33.8%	216
3	65	74.7%	22	25.3%	87
4	31	83.8%	6	16.2%	37
5	1	100.0%	—	0%	1
Total	489	63.8%	277	36.2%	766

ing the years 1974 and 1975 responded, we recorded a response rate of
over 80% for the years 1982, 1983, and 1984 (table 1.6).

In any case, taking into account the size of the questionnaire and the
postal difficulties in many DCs, we can consider the response rate
satisfactory.

Characteristics of the Survey Population

Before presenting and interpreting the results, it seems important to
point to the specific characteristics of the study population in order to
avoid overgeneralizing the conditions of DC scientists. In what way is
our population representative of the DC scientists?

As we just saw, relative to the world distribution of scientists, the
Latin American scientists, and even more so the African scientists, are
overrepresented; and the Asian scientists are underrepresented in our
sample. The main effect of the African overrepresentation is that our
study lends greater importance to scientists working in the scientifically least developed of the DCs.

The scientists in our sample work in fields of research given priority

Table 1.6. Responses to Questionnaire by Year of Approval of First Grant

Yr. of Approval of First Grant	Responses		No Responses		Total
1974	20	46.5%	23	53.5%	43
1975	18	35.3%	33	64.7%	51
1976	44	55.0%	36	45.0%	80
1977	32	56.1%	25	43.9%	57
1978	30	50.0%	30	50.0%	60
1979	43	55.8%	34	44.2%	77
1980	63	65.0%	34	35.0%	97
1981	49	63.6%	28	36.4%	77
1982	62	83.8%	12	16.2%	74
1983	82	87.2%	12	12.8%	94
1984	46	82.1%	10	17.9%	56
Total	489	63.8%	277	36.2%	766

in most DCs, but there are some fields of science and technology that are not covered, in particular the social sciences, medicine, the exact sciences (mathematics and physics), as well as engineering and technology that are not specifically connected to rural development. In other words, the sciences concerned by our study are agriculture, biology and microbiology applied to food and rural development, engineering and technology for rural environments, and, last, a multidisciplinary field that involves a variety of subjects such as chemistry, biology, taxonomy, pharmacy, and, on the periphery of medical sciences, natural products, in particular active substances in medicinal plants.

It is obvious that scientists working on physics would be more inclined toward basic research and maintaining close contacts with the international scientific community than scientists working on agricultural research. But it is difficult to compare the relative importance of IFS-supported fields with general statistics like the ones produced by UNESCO because the definitions of the latter's research areas are usually too broad. The most interesting estimates are probably the ones put forth by C.H. Davis (1983), although they apply only to sub-Saharan Africa. For the period from 1970 to 1979, the ISI records show that publications included 22.3% for agriculture, 22.4% for biology, and 38.2% for medicine. Despite the 22.3% figure for agriculture, the citation record was 14.5%. This can be explained because of the site specificity of research carried out in the agricultural sciences, in the broad sense of the term. According to Garfield (1983), the 14 Third World journals that generate the largest number of citations for Third World authors are devoted to chemistry, biology, and medicine. This justifies the hypothesis that the work carried out in the IFS-supported research fields, except for the chemistry of natural products and microbiology, are underrated in the computations referred to above because they lie somewhat outside the mainstream.

Without being too presumptuous, we can say that the fields studied by scientists in our survey group cover over 50% of the fields that are under study in the DCs. The research our scientists conduct is site-specific and applied rather than basic, although the dividing line between the two is not easy to plot. It would be of dubious value to make comparisons with UNESCO statistics. As an example, it is interesting to see that conclusions from the 1985 Statistical Yearbook show that in 1976 Niger devoted 100% of its research and development activities to basic research!

The percentage of scientists in our sample who work in university or other academic institutions, i.e., over 70%, may at the outset seem excessively high. Taking Davis's (1983) estimates, we see that, at least for sub-Saharan Africa, this figure is perfectly acceptable, since 65% of the total scientific output is produced in universities. Most of the DCs, indeed, do not have an industrial research capacity, except for countries like Brazil, Mexico, India, Argentina, and the Republic of Korea, which, when taken together, produce 60% of the manufactured goods coming from Third World countries (Perrin 1983, p. 13).

What most clearly singles out the survey population may be that it is comprised of internationally selected scientists, chosen according to criteria that have become ever stricter over the years. Since the selection/application rate in recent times has been one out of five, they are no doubt among the most highly qualified scientists of their countries. (Remember that 60% of them have a Ph.D. and that 76% of their degrees were received in the industrialized countries.) Further, the group is composed of scientists who have decided to work in their home country and who, when receiving the IFS grant, were employees of research and/or training institutions where, under very varied conditions, they had the opportunity to conduct research.

What one of the scientists told me rather clearly reflects what many of them feel: "The economic situation in my country makes research very difficult and frustrating, but I decided to stay anyway." Even the ones who temporarily—often for financial reasons—leave their countries are determined to return home at the end of their contracts, even if they have the opportunity to stay abroad. We will come back to this. Potential emigration is partly unfulfilled because most of the scientists are attached to their country and ther home environment.

Bernard Houssay, the Argentinian Nobel Prize recipient, said, "Science does not have a country, but a scientists does. . . . the country where he was born, or raised and educated, the country that gave him a place in his professional career, the country of his friends and family" (cited in CIMT 1970, p. 450).

A further characteristic of this sample population is that its members are mainly at the beginning of their scientific career when they receive the IFS grant. For many of them, this is their first opportunity to conduct independent research. IFS funds enable them to be more flexible, enjoy more credibility in the eyes of their superiors, and gain greater confidence in themselves. The following comment by a Latin American grantee describes the situation well: "As an IFS grantee I have funds available and, for the first time in my career, can use funds as I see fit. I have also been able to have some influence on the choice of research areas within my institution."

Chapter Two

Origins and Education

Social and Family Origin

I come from a rural village in the eastern region of Ghana. My parents are illiterates but they had great interest to educate their children because they felt that education went with good jobs in our society. I am the second born out of a family of ten children. I have four brothers and five sisters. I was motivated to acquire a higher education through a prominent academician from my village. When I was a small boy, any time this man came to our village I would visit him, and take inspiration from him. This man is now a professor of linguistics at a university in Ghana. I however developed love for science and scientific research when I became aware of the numerous problems facing Ghanaian agriculture which needed to be solved.

This young Ghanaian, who now has a doctorate from a prestigious British university, has come a very long way. After getting his degree, and before returning to Ghana, with help from IFS he was able to work at the renowned Rothamsted Experimental Station under the supervision of scientists known the world over.

This career story illustrates how fast a relatively large number of scientists from DCs have been able to advance from a small village to a big city. At the end of this sociointellectual adventure, they go on to become members of the intelligentsia, leaving behind them their home village where they seldom feel they still belong. The change is not always radical. To determine the social and family origin of the scientists in our study population, we used the father's profession as an indicator to tally with the milieu of origin (capital, village, etc.) in developing our analysis. The results are reported in table 2.1. It is interesting to observe that close to one-third of the researchers have a farming background, and one-fifth spent their childhood in a village. Before going any further, we should explain that the category "agriculture" embraces all the agricultural paid laborers and tenant farmers, regardless of the size of the farm, and cannot in any way be said to form a homogeneous social group. The average size of the holdings and the productivity levels differ from country to country. The average size for the DCs as a whole was between 0.5 and 3 hectares per family (Todaro 1977), which supports the thesis that many of them came from small

Table 2.1. Father's Profession as Related to Milieu and Degree of Urbanization

Father's Profession	Degree of urbanization				
	Capital City	+50,000 Inhab. Other Than in the Capital	2,500 to 50,000 Inhab.	−2,500 Inhab. and Villages	Total
Agriculture	12 2.6%	21 4.5%	44 9.4%	62 13.4%	139 29.9%
Liberal professions & senior management	40 8.6%	38 8.2%	26 5.6%	7 1.5%	111 23.9%
Crafts and commerce	24 5.2%	30 6.5%	22 4.7%	13 2.8%	89 19.2%
Middle management	24 5.2%	17 3.7%	7 1.5%	6 1.3%	54 11.6%
Office staff	17 3.6%	15 3.2%	4 0.9%	0 0%	36 7.8%
Laborers	5 1.1%	7 1.5%	4 0.9%	1 0.2%	17 3.7%
Service staff and caretakers	1 0.2%	3 0.6%	0 0%	1 0.2%	5 1.1%
Other categories	4 0.8%	2 0.4%	4 0.9%	3 0.7%	13 2.8%
Total	127 27.3%	133 28.6%	111 23.9%	93 20.2%	464 100.0%

subsistence-level farms. Although the results did not separate the "paid laborers" from the "tenant farmers" category, we ascertained that many paid laborers were farmhands working on large plantations growing export crops.

There are at least two reasons why these preliminary findings hold few surprises. The first is that in Africa and Asia 78% of the population earn their money (or their meals) from subsistence agriculture or as paid agricultural workers. In Latin America the figure is 47% (ILO 1974). The second reason is related to the fact that most fields of research somehow focus on agriculture and rural development. A survey conducted in the United States on 10,000 people who graduated from American universities between 1935 and 1960 showed that 49% of the researchers working on agriculture had fathers who had worked in agriculture, while only 15.6% of the researchers working in other fields of science and technology had this heritage (Harmon 1965). A more recent study carried out in the United States shows the effects of having a family with an agricultural background on the agriculture-related fields of research a scientist selects. "Although the U.S. farm population is now less than 4% of the total population, 38% of all agricultural scientists come from farm backgrounds. In particular in agronomy, animal science . . . scientists typically have farm backgrounds" (Busch and Lacy 1983 p. 54).

The work by Bourdieu is most informative, notwithstanding the fact that it applies to a French student population and research scientists. He reports that for the 1961–62 school year, 28% of the students in national agricultural schools of France came from farm backgrounds, while the average for all educational disciplines taken together was 6% (Bourdieu and Passeron 1964, pp. 20–21). This said, we still see that most of the DC scientists who come from farming families have a much more modest social background than their colleagues from the developed countries. For this category we can unhesitantly use the term *social advancement*.

For the other categories, we have to make allowances for the existence of unequal opportunities. Results unquestionably prove that the grade school and then the university system have selection criteria that are hardest on the least favored classes without, however, totally excluding them. The intermediate categories (especially crafts and commerce) are in a rather good position (19.2%). Here again, the results must be interpreted with great caution. The subcategory called "commerce" includes a heterogeneous range that stretches from the little stalls in the marketplace to large import-export firms. Our personal contacts with the grantees of our study population and the answers to the questionnaire support the assertion that the scientists whose parents are artisans or work in commerce come from relatively modest families. The percentage of grantees whose fathers were in the category of "office staff" was the same, i.e., ±8%, as Bourdieu found for students in French universities, all disciplines combined, for the 1961–62 school year (Bourdieu and Passeron 1964, pp. 20–21). The percentage of sons and daughters of "laborers" (3.7%) was lower, but

Table 2.2. Geographical Breakdown of Researchers by Sex

	Men		Women		Total
Asia and Pacific	170	77.0%	51	23.0%	221
Africa	165	91.0%	17	9.0%	182
Latin America	73	85.0%	13	15.0%	86
Total	408	83.4%	81	16.6%	489

this can easily be explained by the lower rate of industrialization in the DCs.

The high percentage of researchers (23.9%) whose parents are in liberal professions or senior management positions, a social category that comprises a very small percentage of the population in most DCs, confirms the inequality of opportunity.

Eisemon provided further evidence for these results through interviews he made in Kenya and in Nigeria in 1978: "African scientists, like most other Africans with higher education, are usually the first in their families to receive secondary and higher education. Many, particularly in Kenya, come from rural backgrounds" (Eisemon 1979, p. 512).

Breakdown by Gender

Figures in table 2.2 show that only 16.6% of the IFS grantees are women, who thus, as a group, are underrepresented. However, a quick comparison with the situation in the developed countries of the world mellows our initial reaction. For example, in 1982 only 13% of the scientists and engineers in the United States were women, and this was a 200% increase over the 1972 figure. Women, however, make up 45% of the work force in that country (National Science Foundation 1984, p. 1).

In a country like Sweden, which is well known for its efforts in favor of equality of the sexes (Gaillard 1983), in 1982 women accounted for only 12% of the research scientists (SOU 1983). Furthermore, as the academic level rises, the number of women at each level declines; over half the student body in the high schools is composed of girls, but only 25% of the Ph.D. candidates and 3% of the university teachers are women (SOU 1983).

Using mean percentages of women and grouping the countries in our study obscures regional disparities and important differences between the countries. Women researchers in our population figure as follows: 9% for Africa, 15% for Latin America, and 23% for Asia. The

Table 2.3. Gender of IFS Grantees by Research Area

Research Area	Male		Female		Total
Food science	33	62.0%	20	38.0%	53
Forestry	30	77.0%	9	23.0%	39
Natural products	72	80.0%	18	20.0%	90
Aquaculture	83	85.0%	15	15.0%	98
Crop science	100	88.5%	13	11.5%	113
Rural technology	19	90.5%	2	9.5%	21
Animal production	71	95.0%	4	5.0%	75
Total	408	83.4%	81	16.6%	489

Philippines (36%) and Thailand (33%) had the highest percentages; there were no women researchers in the Republic of Korea and only 10% in Sri Lanka. Some African countries such as Tunisia (27%) and Tanzania (23%) have a laudably high percentage compared to the continent as a whole, while countries like Burkina Faso, Morocco, and Senegal rank far below the average.

Sample comparisons, excluding the countries with very few grantees, confirm that the percentage of women in the IFS population coincides rather closely with the national percentage in the countries we studied. Thus, according to a study carried out by the International Service for National Agricultural Research (ISNAR) in 1984 on 1,400 researchers and technicians of the Department of Agriculture in Thailand, 38% were women (Elliot 1984). This study also brought out a strong degree of disciplinary specialization; women tend to choose disciplines that involve laboratory work and that offer jobs in the capital (Elliot 1984, p. 3). Women exceed their overall average in the fields of food science, forestry, and natural products. This is particularly true in Asia, where the percentage of women in research is the highest.

As for Senegal, in 1984 we found that of the ±400 scientists and technicians working at the Institut Sénégalais de Recherche Agronomique (ISRA), only 4.5% were women. Statistics on the 1982–83 teaching staff at the University of Dakar showed a female component of 8.5% (over half were expatriate women). Some 7% worked in the faculty of Medicine and Pharmacy and the Faculty of Arts and Humanities. Going back to 1980, in Sudan, out of 123 agricultural researchers only 4 (i.e., 3%) were women. They worked in food sciences and in nutrition; there were no women in the other disciplines (Lacy et al. 1983, p. 19).

In food science, research by women in our sample bears directly on

nutrition and the production of foodstuffs: the improvement of traditional procedures to ferment foods, the production of microbial pro-
teins, and studies linked to the contamination of foods by mycotoxins
and aflatoxins. In forestry women mainly research mycorrhizal associations, or the isolation, determination, and culture of mycorrhizal
strains in the laboratory, or some field of taxonomy or ecology. Although some women run field experiments, forestry research demanding extended assignments away from home is left to the men.
Natural products, mostly studied in the laboratory, and aquaculture
constitute disciplines at the crossroad; but here again it is interesting to
note that women researchers in this field concentrate on nutrition and
the development of foods for various aquatic organisms, on the development of vaccines, and on the parasitology of fish diseases. The three
remaining fields are typically the province of men and often require
temporary or permanent posting to isolated research units outside the
capital or a large town.

Women decline to live outside urban areas not only because of their
discipline. Other factors such as the marital status, the number of
children to support, and the spouse's profession can also affect the
researcher's region of residence and, subsequently, his/her method of
research.

Marital Status and Spouse's Profession

Compared to the national trend, scientists in our study population
enter marriage late, since 70% in the 25 to 29 year age group are unwed,
as are close to one-third in the 30 to 34 year age group, and one-fifth in
the 35 to 39 year age group. One reason may be that many of them had
long years of schooling and extended journeys abroad. Another reason
may be that many students (85%) were in contact with the Western
model during their studies outside of their home countries. (For more
details on the marital status, see table A1, Appendix C.)

The Western standard also seems to have been adopted for the
number of children, since two-thirds of our scientists have at most two
children. The maximum was eight children; the father was an African
scientist over 50 years of age. Only 4% of the scientists have more than
four children. Finally, considering the galloping population increases
in the countries of our study, it was interesting to note that close to half
the scientists in the 30 to 34 year age group and over one-fourth in the
35 to 39 year age group had no children at all. (For more details on the
number of children per family, see table A2, Appendix C.)

Who do our scientists marry (table 2.4)? There is a strong endog-

Table 2.4. Profession of Spouse, by Gender

Profession of Spouse	Male Grantees		Female Grantees		Total	
No profession	109	31%	1	2%	110	27%
Liberal professions & senior management						
Scientists	55	16%	26	46%	81	20%
Professors	67	19%	6	11%	73	18%
Liberal professions	14	4%	5	9%	19	5%
Senior administrative	11	3%	2	3%	13	3%
Engineers	5	2%	7	12%	12	3%
Subtotal	152	44%	46	81%	198	49%
Middle management						
Medical social prof.	14	4%	1	2%	15	4%
Civil servant	13	4%	4	7%	17	4%
Technical staff	9	2%	2	3%	11	3%
Subtotal	36	10%	7	12%	43	11%
Office staff	33	10%	0	0%	33	8%
Industry and commerce	6	2%	3	5%	9	2%
Other	11	3%	0	0%	11	3%
Total	347	100%	57	100%	404	100%

amous trend since half of the spouses are scientists and teachers. The marriage strategy (late marriage, strong endogamy, Malthusian behavior) seems to characterize a very rational approach to reproduction for this emerging intellectual class. Under the influence of the Western model, which purports that small families are more mobile and get along better socially than large families, the scientists produce as many children as they think they can establish at a level they would be satisfied to occupy themselves. The investment required for research quite clearly implies postponing marriage and the first child. Since in research the social status that accompanies the profession seems to take more time to acquire than in other professions, scientists have to—and seem to be prepared to make—the relevant sacrifices.

This endogamous trend seems to be stronger among the women researchers than among the men, since close to half of them (45%) married research scientists. This provides us with an additional explanation for the fact that women, more than men, prefer exercising their profession in the capital or in a large town.

We should also note that more than two-thirds of the wives of researchers have their own professional activity, which reflects a substantial social and mental break from their original social milieu. Finally, none of the scientists' spouses worked in farming, and very few (2%) worked in industry and commerce, while one-fifth of their fathers belonged to this social category.

Our efforts to find a link between the length of residency in a Western country and the adoption of the Western model as a system of reference led us to the conclusion that there was no clear-cut cause-and-effect relationship. The number of years spent abroad did not seem to have any influence on the profession of the spouse or the number of children. Actually, we found that people who had spent the most years in Western countries (over 10) tended to have the most children (table A3, Appendix C). But we have to remember that these are our oldest scientists and tended to have four or more children. Adopting the Western model seemed to stem less often from acculturation caused by an extended stay in the West than from the scientist's profession. Constraints linked to aggregated effects of intellectual advancement (time) are probably the most decisive.

Similarly, there does not seem to be any very significant relation between the number of years spent abroad and the social origin of the scientists except, to a certain extent, for the "crafts and commerce" and the "managerial staff" categories whose scientists tend more to study abroad for extended periods of time (see table A4, Appendix C). Let us look at the training component. Where, how, and for what degrees did our scientists study?

Training

"I was born in Fiji and attended primary and secondary school there. The secondary school was the Marist Brothers High School in Suva where I obtained my Fiji Junior Certificate, Overseas Cambridge School Certificate, and New Zealand University Entrance Certificate. . . . I wanted to get into a university to do further studies. At that time there was no university in Fiji so I went to the University of Canterbury in New Zealand in 1964. I was at this university for five years obtaining my Bachelor of Science and Master of Science degrees." This Fiji scientist went on to study at the University of Queensland in Australia, where he received his Ph.D. in rural economics in 1976. In the meantime, at the end of the 1960s the University of the South-Pacific was created in Suva. In sum, he went abroad for studies three times, and it took him over nine years abroad to obtain his Ph.D. (He interrupted his studies to work temporarily at the Ministry of Agriculture in Fiji.)

Until relatively recently, except for certain countries such as India, many DC students had to leave their home countries to attend a university and obtain the education needed to become scientists. Studying abroad is nothing new and is not specific to young people from DCs, but it is noteworthy that the percentage of students from DCs in the total foreign student bodies has increased considerably in most Western countries since the 1960s. In the United States, for instance, the figure rose from 34,232 in 1954–55 to 235,509 in 1977–78. During that same period, the Africans, as a percentage of the total foreign student figures, rose from 3.4% to 12.6%, while the number of Europeans and Canadians fell from 28.9% to 13.7% (Maliyamkono and Wells 1980, p. 1).

During the colonial era, most of the (very few) students who were sent abroad for their education studied in the colonizers' country. During the preindependence years, increasing numbers of DC students applied to study abroad, and the number of scholarships made available by industrialized countries went up considerably. This evidenced increased awareness of the importance of education and the role of higher education in development-oriented science; it also reflected the donor countries' desire to maintain—or acquire—political and economic influence in newly independent states (Gaillard 1985a, p. 94).

At the time of independence, there were some universities in the DCs, but they did not go as far as the doctoral level and did not offer a full range of science and technology courses. In Latin America, for instance, during the colonial era there were 23 universities but only 150 alumni by the end of the eighteenth century (Botelho 1983, p. 16). In 1857 the British colonizers created the frst universities in Asia, Calcutta, Madras, and Bombay. The University of Cairo was founded in 1908. The first universities of Black Africa are far more recent. The first courses to be given in University College, Ibadan, Nigeria, started in 1948; and the first science majors graduated in 1950 (Kolinsky 1985, p. 34). The University of Dakar, the oldest university in French-speaking Black Africa, was started in 1957 and became Senegalese in 1960 (Gaillard 1985b, p. 6).

In some countries, after the first university was created, change took hold very quickly, especially in the 1960s. In present-day Brazil there are 60 universities plus 800 other institutions of higher learning, while before 1965 a university education was available—to a very limited number of students—almost exclusively at the University of Sao Paulo. The number of students has made a sharp upward turn, going from 200,000 in 1968 to over 1.1 million in 1977. Brazil now offers close to 600 graduate programs in some 30 independent institutions and universities. Two-thirds lead to a master's degree, one-third to a doctorate. The University of Sao Paulo in 1977 offered 100 master's programs and 66 doctoral programs in a great variety of disciplines.

Brazilian officials query the quality of these programs and feel that only one-third of the doctoral programs are academically valid and that more than half of these are offered at the University of Sao Paulo (Schwartzman 1978, p. 545). This explains why the Brazilian National Council for Scientific and Technical Development (Conselho nacional de desenvolvimento cientico e tecnologico, CNDCT) granted 1,000 scholarships to Brazilian students in 1984 to receive doctoral or postdoctoral training mainly in the United States, Great Britain, France, Belgium, Canada, Federal Republic of Germany, Australia, Sweden, and Spain (CNDCT, personal communication). This is only the tip of the iceberg because preparatory studies for a career in scientific research form a subject that the DCs find difficult to harness, and training abroad is more often than not the result of the student's personal initiative rather than a well-orchestrated government plan. Since the beginning of the 1960s, the spectrum of countries that receive these students has grown considerably, although the DCs, regardless of the postcolonial political vicissitudes, are still marked by their colonial heritages.

In 1974, pursuant to major political upheavals, nearly the whole French scientific community of Madagascar packed up and left as the country went on to conclude numerous technical assistance agreements with COMECON countries (Gaillard 1984b, p. 2). An analysis of Malagasy students studying abroad between 1975 and 1982 shows that out of 2,000 students half studied in the USSR and (despite the above) 20% in France; of the rest 14% studied in Romania, 6% in Algeria, 3% in Cuba, 1% in the United States, and 6% elsewhere. Further, in a summary survey we made in 1984, out of 69 scientists working at CENRADERU (the national center for research applied to rural development), we discovered that half of our respondents had studied exclusively in Madagascar, and 37% had studied in France. It should be noted, however, that only 20% of them had a doctorate or a Ph.D., that 65% were between 30 and 39 years of age, and that only 10% were under 30 years old. At the University of Madagascar, the fact that a relatively large number of researchers, including many from the youngest age bracket, had obtained their SBA (diploma for advanced studies in applied biological sciences) is a clear indication of the Malagasy government's determination to offer training in research within the country.

The Brazilian and Malagasy examples touched upon above show us that training in research is undergoing rapid change with an expanding host of countries offering study opportunities. We were also able to see that speed is not always compatible with quality and that at the doctoral level there is still strong reliance on the industrialized countries. Let us return to our study population.

As we saw in the introduction, over 60% of the scientists in our population have a state or third-cycle doctorate, Ph.D., or a doctor of science degree. This high percentage is the result of selection and is not the norm in the DCs. A study conducted on 20,000 scientists in 32 countries, including 4 African countries, showed that in the 1960s only 9% were in the highest educational group, i.e., had a doctorate, 27% had a master's, and 64% had a degree equivalent to the "licence" or a bachelor's (Oram and Bindlish 1981, p. 5). Actually, in many DCs research is entrusted to scientific personnel of the "bachelor's" or "licence" level. The study also indicated that the 4 African countries (Kenya, Nigeria, Madagascar, and Sudan) on the average had a higher percentage of scientists with a third-cycle (postgraduate) diploma than in Asia or Latin America; the average for these African countries was that 27% of the scientists had a Ph.D. and 37% had a master's.

These statistics must be applied with great caution. Despite considerable efforts, there are still few DCs that have a very accurate picture of their scientific and technical potential. For Sudan, Lacy et al. (1983, p. 23) purport percentages that are even higher than the ones that describe the researchers in the Sudanese Agricultural Research Corporation (ARC): out of 161 scientists working for ARC in 1981, 50% had a Ph.D. This percentage is even higher (65%) if we exclude the research assistants. Further, 12 out of the 13 scientists who had been seconded elsewhere that year also had a Ph.D. Finally, 96% of the Sudanese Ph.D.s had received their diplomas, in equal proportions, from American and from British universities (Lacy et al. 1983, p. 24).

Table 2.5 shows that three-fourths (76%) of our study population did their doctoral studies in the industrialized countries. There is no significant difference on the basis of country of origin. A country analysis, however, shows that doctorates obtained in Asia and in Africa were dispensed in a very limited number of countries. In Asia, 22 (60%) of the 37 students who obtained their doctoral degree studied in India. In Africa, 11 (40%) out of 28 studied in Nigeria, mainly at the University of Ibadan.

Table 2.5. Doctoral Studies: Continent of Origin and Continent of Studies

Place of Studies	Continent of Origin							
	Asia and Pacific		Africa		Latin America		Total	
Industrialized countries	107	74.3%	91	76.5%	29	80.6%	227	75.9%
Asia	37	25.7%	0	0%	0	0%	37	12.4%
Africa	0	0.0%	28	23.5%	0	0.0%	28	9.4%
Latin America	0	0.0%	0	0.0%	7	19.4%	7	2.3%
Total	144	100.0%	119	100.0%	36	100.0%	299	100.0%

Table 2.6. Place of Doctoral Degree by Date of Studies

Dates of Study	DCs		Industrialized Countries		Total	
1980–85	30	31.3%	66	68.7%	96	32.4%
1975–79	20	22.2%	70	77.8%	90	30.4%
1970–74	7	10.0%	63	90.0%	70	23.7%
1960–69	13	32.5%	27	67.5%	40	13.5%
Total	70	23.6%	226	76.4%	296	100%

The percentage of doctorates obtained in the DCs has been increasing steadily since the early 1960s and rose from 10% in the 1970–74 period to over 30% in the 1980–85 period, as table 2.6 shows. For the period 1960–69, out of 13 doctoral degrees obtained in the DCs, 7 were awarded in India, 2 in Nigeria, and the other 4 in Argentina, Brazil, Egypt, and Pakistan. In percentage terms the number of doctorates obtained in the DCs was rather high during this period partly because the first wave of DC students who graduated in industrialized countries returned home at the end of the 1960s and even more returned during the early 1970s. This is also an immediate result of the first contacts established by the IFS secretariat staff and scientific advisers in the above mentioned countries (except Brazil). In January 1975 alone, there were three missions to India, and most of the Indian grantees

Table 2.7. Number of Ph.D.s by Number of Years of Study Abroad

Years Abroad	Researchers with Doctoral Degree		Total Population		% of Researchers with Doctorate/ Total Population
0	20	6.7%	73	15.2%	27.4%
1–2	35	11.8%	113	23.5%	31.0%
3–4	89	30.1%	118	24.3%	75.4%
5–9	126	42.6%	151	31.4%	83.4%
10–20	26	8.8%	27	5.6%	96.3%
Total	296	100.0%	482	100.0%	61.4%

were selected in 1974–75 from among the young scientists who received their Ph.D. in the 1960s. Afterward IFS adopted a much more discreet policy toward India in an attempt to stem the flow of applications from that country.

There is still strong reliance on studies abroad for research scientists. Table 2.7 shows that over 60% of our population studied abroad for over three years in order to obtain a doctorate. It also shows a correlation between the (increasing) number of years abroad and the (growing) probability of obtaining a doctorate. In other words, the probability that scientists who spend ten years or more abroad have a doctorate is 3.5 times greater than for scientists who studied exclusively in their home country.

Dependence on other countries for education is also directly related to the level of diploma being sought; the more advanced the degree, the greater the dependence. Hence, for the doctorate 75% of our population studied abroad, for the master's degree 45%, and for the bachelor's degree 10%. Most students who went abroad for the bachelor's degree came from Africa and the Pacific, but there were also some from Asian countries such as Malaysia and Thailand that offer their own bachelor's program.

Although we have no statistical proof, we have noticed that a student who has the choice between studying at home or abroad will generally choose the latter. Besides the economic benefits that accompany a stay in a foreign country, a diploma obtained in an industrialized country is usually rated higher than a diploma from a DC. Moreover, in the diploma bargain market, there are hierarchies and clans! In France alumni of certain universities are considered to control certain institu-

Table 2.8. Study for Master's Degree: Continent of Origin by
Continent of Education

M.S. Preparation	Continent of Origin							
	Asia and Pacific		Africa		Latin America		Total	
Industrialized countries	58	35.6%	64	56.6%	23	50.0%	145	45.0%
Asia and Pacific	105	64.4%	4	3.6%	3	6.5%	112	34.8%
Africa	0	0.0%	45	39.8%	0	0.0%	45	14.0%
Latin America	0	0.0%	0	0.0%	20	43.5%	20	6.2%
Total	163	100.0%	113	100.0%	46	100.0%	322	100.0%

tions, just as the "Chicago Boys" might be known for their political and economic influence over the government of Chile.

We have seen that for the doctoral degree, reliance on foreign universities seems to be much the same regardless of continent of origin. This does not hold for the master's and the bachelor's; Africa is the least independent. Table 2.8 shows that for the master's degree, some 65% of the Asian researchers stayed in Asia, as against 40% of the African students who stayed in Africa. All the Indian researchers, plus three researchers working in Africa and three working in Latin America, obtained their master's degrees in India. Actually, four of the latter six were of Indian origin and worked in Panama, Mauritius, Brazil, and Morocco. There was also a Pakistani who completed his master's degree in Pakistan and is now working in Uganda.

India is also the country with the highest percentage of Ph.D.s, since 25 out of 26 scientists in our population have reached this qualification. It is the country that is the least dependent on foreign universities for this level of education: 21 of the 26 graduated in India and 4 in the United States. This confirms our thesis that India is the DC that is the least dependent on foreign universities at the diploma level.

Among the countries that train DC students to the doctoral level, there are three that stand out on the international scene since they alone provided 80% of the doctorates obtained abroad by the IFS grantees; viz., United States (34%), Great Britain (26%), and France (20%), followed rather far behind by Australia (6%) and Canada (4%). Table 2.9 shows the number of students per DC country and per country of education.

Table 2.9. Industrialized Country of Study by Home Continent

Country of Ph.D.	Home Continent								
	Asia and Pacific		French-speaking Africa		English-speaking Africa		Latin America		Total
United States	40	51.3%	5	6.4%	21	26.9%	12	15.4%	78
Great Britain	35	59.3%	4	6.8%	15	25.4%	5	8.5%	59
France	2	4.4%	39	84.8%	0	—	5	10.8%	46
Australia	12	85.7%	0	—	1	7.15%	1	7.15%	14
Canada	7	70.0%	1	10.0%	1	10.0%	1	10.0%	10
Total	96	46.4%	49	23.7%	38	18.3%	24	11.6%	207

One-third of the students who sought education abroad obtained their doctorate in the United States, a country chosen mainly by students from Asia and English-speaking Africa but also from Latin America and, to a lesser extent, French-speaking Africa. In 1970 66% of all foreign Ph.D. students in the United States came from the DCs; by 1980 the figure had risen to 77% while the percentage of students from the rich countries during that same period had dropped from 38% to 23% (NSF 1981).

In Great Britain the foreign student component of the total student body was 19% in 1962; the figure rose to 35% by 1976, 43% of whom came from commonwealth countries. The geographical distribution of students in our population who obtained their doctorate in Great Britain and in the United States is quite similar, except as concerns the students from Latin America who more readily went to their northern neighbor for their studies.

Students from French-speaking Africa clearly prefer studying in France, since 85% of the doctorates completed in France were dispensed to students from this region. Nearly all the foreign students studying in Australia came from Southeast Asia and the Pacific. Most of the students who complete their doctorate in Canada come from Asia.

Considering the small numbers involved, it would be unscientific to generalize the results reported above, especially for doctorates obtained in the last two countries. Yet the students' home countries and the countries they choose for their Ph.D. studies rather closely coincide with international geopolitical zones of influence, even though our analyses are based on results derived from a small group. It is illustrative that the four researchers in our population who completed their doctorate in Japan came from three Southeast Asian countries and from South Korea. The USSR constitutes a glaring gap on this educational map, having conferred only 2 doctoral degrees (to a Vietnamese and a Sri Lankan) out of the 226 obtained in foreign universities.

Although the present-day tendency is to diversify sources of education (as we will see later with Morocco), some DCs are still determined to send their students to certain specific countries, e.g., out of 14 Ph.D. students from the Philippines, 11 studied in the United States. Another example was Sri Lanka; out of 21 Ph.D. students, 17 studied in Great Britain and 4 in the United States. On the other hand, a country like Thailand, which has never been colonized, tends to assign its Ph.D. students more evenly since 7 went to the United States and 9 to Great Britain and also approaches other countries, since 2 students went to Australia, 2 to the Federal Republic of Germany, and 1 to the Philippines.

The countries of study also have their preferred disciplines. Three of the four doctorates awarded in Japan were in the field of aquaculture, a field that was also favored in Canada. Although the United States offers courses in all fields, it has special strength in the chemistry of natural products and in crop sciences, especially in cropping methods and plant improvement. Over half the doctorates completed in Australia focused on crop sciences with a bias toward plant physiology, mainly fodder crops. Here again we must avoid overgeneralizing lest we err, but this is a pertinent line of research that requires more thorough analysis and probably a bigger population in order to produce more definitive conclusions.

In this phase of "construction," training abroad is important for the DC scientific communities but must be considered as supplementary to education within the home country. Under no circumstances should it be used as a replacement for education within the DCs, which, during this transitional phase, must endeavor to meet their national requirements for scientific and technical staff through an intelligent balance between these national and foreign sources of education.

Furthermore, doctoral and postdoctoral studies in developed countries supported by the host countries are of considerable help to the DCs in strengthening their national research capability at relatively little expense: "Postdoctoral appointments in the United States have become less a means of enriching the research capacity of a few outstanding individuals in a few centres, and more a means of increasing the size of the staff available to work as members of a research team in the physical and biological services" (Kidd 1983, p. 409). We think that this remark could also be applied to doctoral courses and, in this case, applies to a far larger group of people.

The current tendency is toward a mixture, sometimes called a "sandwich," that alternates fieldwork and collecting data in the student's home country with lecture courses and thesis preparation (and defense) in a foreign, industrialized country. This is a system that has been adopted by many institutions of higher learning in the DCs, e.g., the Institut Agronomique et Vétérinaire Hassan II (IAV) in Rabat, Morocco. The "sandwich" formula is included in a global education policy designed to diversify sources of education that traditionally were limited to France and to coordinate training dispensed by virtue of agreements concluded with various foreign countries and institutions.

For the third-cycle (postgraduate) doctorate in agronomy, most of the students are sent abroad to France, the United States, Belgium, Great Britain, Canada, etc., for their fifth year and then return to their institute and spend most of their sixth year preparing a thesis. This is a rather recent pattern, and many Moroccan students seem to prefer applying for a scholarship and doing all of their studies abroad. Quite interestingly, out of 19 Moroccans who applied for an IFS grant in 1982, 12 had completed part or all of their higher education in France, 4 at IAV in Morocco, 2 in Great Britain, and 1 in Switzerland. A few had attended advanced training courses in France, the United States, Germany, Great Britain, India, and Sweden (Gaillard 1982).

This academic diversity provides a well of ideas and inspiration for IAV to draw from in developing innovative methods for research training. An eclectic approach such as this can also cause confusion and make it difficult to evaluate foreign training schemes and diplomas, especially when a thesis has been written in a foreign language that is incomprehensible to nearly all the scientists and administrators in the student's home country. The problem is further aggravated when a group of countries decides to organize a system of higher education at the regional level, as is the case in the Pacific: "The diversity of colonial experiences throughout the Pacific has resulted in a plethora of educational systems, with students studying in different languages, following different curricula" (Pernetta and Hill 1984, p. 164).

Postgraduate programs are gradually being offered in the DCs. This has many advantages, e.g., students do not have to be sent abroad for extended periods of time and at great cost and can be oriented to topics of direct concern to their country's development. But there is the risk that, at least during the start-up phases, the level of instruction may not be up to the level offered in universities in developed countries. Very often support is provided by scientists from these universities who give courses that local scientists cannot give. This is the case for the diploma of intensified studies (the DEA) in applied biological sciences at the University of Madagascar. Close to 50 students were offered this curriculum during the first two school years; and according to some French university teachers who worked on the program, it is just about as good as many French DEAs (Gaillard 1982, p. 2).

Not all the postdoctoral students in DCs, whether they studied abroad or in their home countries, are slated to become research scientists—far from it. What persuaded the scientists in our study population to make this choice? How do they choose their first research topic? What are the most decisive criteria, and how do they differ from topics chosen by researchers in the developed countries? That will be the subject of the next chapter.

Chapter Three

Research As a Profession

The Choice

Research scientists in DCs long to have a proper professional status; often draft texts have been prepared and then stored away in anticipation of better times to come. It is not up to us to discuss whether research science is, or is not, a profession in the DCs. "Research is often carried out as part of some profession or system designed to uphold professional standards or value systems that are not specific to research" (Barel and Malein 1973, p. 933). This was the case for a Moroccan scientist who had to bend his institution's rules in order to obtain a research budget. His problem was not at all an uncooperative boss or even fear of outside influence but merely that the statutes of his institution had no provisions for research in its normal program of activities. We can also look into the ambiguous status of 70% of the scientists in our study who were teachers-cum-researchers but whose primary occupation, and in some cases, whose only recognized occupation, was teaching. What Jean Perrin said about the French universities in 1933 could also apply to certain DC universities now: "Allocating funds to scientific research in a university is an irregularity to which the Administration agrees to close its eyes" (cited by Salomon 1970, p. 61).

Whether it is irregular or not, more and more institutions are hiring scientists who have been trained in research and who devote varying proportions of their time to research activities. How does one decide to become a researcher in a DC where the education system is often not designed to offer proper preparation for a scientific career and where research, as we will see in chapter 4, is often very difficult to carry out? Obviously, there is no single answer to this question since the choice depends on individual hopes and behaviors, on the social status or even the degree of prestige, and on many other factors that accompany the profession in many DCs.

Marcel Roche, who was part and party to the birth and growth of many scientific institutions in his home country, Venezuela, quite rightly pointed out in 1966 that "the past does not give us any example worth following, and research scientists with the right experience to show the way are few and far between" (Roche 1966, p. 59). It is certainly not mere serendipity that explains the decision to award the Nobel Prize to brilliant research scientists such as C.V. Raman in India and Bernard Houssay in Argentina; and it seems reasonable to expect them to serve as an example for attracting young people to research, even in disciplines other than their own.

In most DCs research scientists do not have a very high social standing or much prestige. Doctors and lawyers and other professionals of that level with the same amount of education as the research scientists are not only better paid but also enjoy a much higher social status. Speaking about Venezuela in the 1960s, Roche said, "I know many examples of young people whose rich parents forbade them to major in sciences or to devote themselves to research often because of the low salaries or uncertain career opportunities. The bourgeois attitude to careers in science is much the same as the attitude to professions in the arts; success is reserved to very outstanding people alone, all the others being condemned to a Bohemian life of uncertainty. The situation has probably changed since the 'Sputnik' was invented, but research is still not seen as a fully acceptable profession" (Roche 1966, p. 57).

The parents of one of our Moroccan friends, who graduated with distinction as a veterinarian in France and recently got a doctorate in animal physiology from the University of Upsala in Sweden, do not understand, or are having difficulty accepting, that their son decided to become a research scientist in Rabat rather than becoming a practicing veterinarian somewhere in Morocco or even in France. This scientists comes from a comfortable family that is in business. The same reaction is also observed in the less favored social classes whose unclear understanding of scientific progress prevents them from seeing the use of research in solving social and economic problems.

How attractive research as a profession can be depends on the country. According to Eisemon, in Kenya scientists have enjoyed a place of special pride in society since the European colonization period when close relationships were established between the scientific and the politico-economic circles. "A scientific career brings an individual into proximity with the commercial and political elite of that [Kenyan] society, [even] if it does not always convey full membership" (Eisemon 1982, p. 137). Eisemon felt that research was appealing not only for professional reasons: "Academics are well remunerated [albeit increasingly less so in comparison to colleagues in the private sector]. Professional conditions at the University of Nairobi are often superior to those available in other kinds of institutions. Academic work is largely unsupervised. Academics enjoy considerable independence in an atmosphere of professional and public trust. Equally important, selection of an academic career does not necessitate a sacrifice of other career opportunities" (Eisemon 1982, pp. 137–38).

In India, ever since the British colonizers brought Western science to Bengal in the nineteenth century, the scientific community seems to have been dominated by the upper Hindu castes, especially the Brahmans. Kapil Raj describes how the Brahmans in their own way "appropriated" occidental ideas and science to give credence to their new—dominant—status in the Indian society (Raj 1986). A glance at the names of the IFS grantees in India and a recent study of scientific personnel in five institutes in Calcutta bears out the continued predominance of Brahmans and the upper Hindu castes in the Indian scientific community. According to this study (Surajit Sinka 1970), out of 386 scientists in the sample only one was Moslem, 83% belonged to upper Hindu castes, even Brahmans, and only 3.3% came from the lower castes. Paradoxically, there is not much prestige attached to the profession of research scientist; and, like most intellectual positions in the public sector in India, it is poorly paid. Eisemon (1982, p. 138) explained it as follows: "The primary attraction of academic work [in India] is not to be found in autonomy, possibilities for professional fulfillment, social prestige, or material reward, but in security of employment."

In an effort to better understand the professional choice made by the research scientists in our population, we asked them to rate eight criteria from 1 (essential) to 5 (not important at all). We first calculated an overall mean for each criterion in an effort to rank the criteria by order of choice for the population as a whole (table 3.1).

Regardless of field or country of research, intellectual stimulation is by far the prime criterion since over 90% of the scientists felt that it was either essential or very important, as we see in greater detail from the following percentages: essential, 65.44%; very important, 27.81%; moderately important, 2.86%; not very important, 1.22%; not important at all, 0.81%.

Social utility was rated second; 60% of the grantees felt that it was essential or very important. Usefulness has always been a motive that research scientists bring out, almost as if it were their justification before society. Social utility is given paramount emphasis in the DCs where research must serve development and all too often is considered to be a remedy for every possible economic and social ill.

The other six criteria were considered moderately important or not very important in making the professional choice. Job security was

Table 3.1. Choosing Research Science as a Profession: Importance of Criteria, in Descending Order

Criteria	Means	Rating
Intellectual stimulation	1.41	1
Social utility	2.18	2
Job security	2.98	3
Promotion prospects	2.99	4
Influence of a professor	3.15	5
Social status	3.25	6
Remuneration	3.33	7
Parental influence	3.89	8

*Means are based on a five-step scale as follows: 1 = essential, 2 = very important, 3 = moderately important, 4 = not very important, 5 = not important at all.
Source: Questionnaire for IFS grantees.

considered moderately or very important for over half the respondents. In countries where researchers are civil servants, such as India and Mali, job security tends to be rated "essential." Prospects for quick promotions are good, but promotions are not always accompanied by higher pay, which explains why these two criteria are rated "moderately important."

Social status is also ranked rather low. The replies do not seem to have any correlation with the socioprofessional origin of the scientists' parents, which tends to confirm that the profession does not command great social esteem.

Finally, parents seem to have very little influence, at least positive influence, on their children's decision to enter research since over 60% feel that this criterion was not very important (20%) or not important at all (41%).

The respondents were free to add other criteria, but very few did so. Actually, an *a priori,* well-weighted choice seemed to explain a career in scientific research after higher education less than the fact that students were selected or had access to a scholarship at the right stage of their education, even when it meant studying subjects that initially did not interest them. This is at least what comes out of the interviews we have made.

Initially I wanted to do medicine but did not get good marks in my qualifying exams in science subjects. This was because in school I did not take any science subjects, but the college I went to agreed to give me a trial run at physics and chemistry. I just got pass marks in these two subjects and thus could not get into medical school. I decided to do botany and zoology for a B.S. I then did an M.S. in plant pathology. While on the M.S. program I got the Rotary International for Postgraduate Study. I thus went to Hawaii University to do tropical plant pathology. I did courses in nematology, bacteriology, virology, advanced plant pathology, etc. On my return to Zambia, I joined the plant protection section where I concentrated on plant diseases, all the while further specializing in nematology. I was, and still am, the only specialist in this field in my country.

I could have done all my studies in Nigeria, but the year before I entered the university U.S.AID had set up a scholarship program for the best students. After having been selected, I went to the United States and rather per chance ended up studying botany, zoology, and chemistry since that was what I was offered. Since I was the top student in my graduating class after obtaining my B.S., it was easy to get a scholarship to go on for a Ph.D. At that stage I was asked to choose between two majors. I chose forestry without really knowing what effect that would have on my future career when I got back to Nigeria. Since at that time I was at the Davis Campus of the University of California, which specialized more in straight agriculture, I was supposed to stay there for two years and then go back to Berkeley to pursue my major in forestry. But finally I ended up staying at the Davis Campus. I had favorably impressed the staff at the Department of Agronomy and was asked to stay. I graduated with a Ph.D., top student in the Department of Agronomy and . . . was asked to stay in the U.S., but six months later I decided to return to Nigeria. Although my country was in the middle of a civil war, I had the feeling that if I stayed in the U.S. I would somehow be betraying my country and my family that I loved very much.

The second interview, which I have shortened, could be summarized by one sentence that came out during our talk: "The opportunity was there." Upon his return from the United States, this Nigerian spent four hard months seeking employment and finally found a job at the University of Ifé. Now he is dean of the Faculty of Agriculture and Agricultural Technology at the newly created Owerri Federal University of Technology.

I want to conclude this series of interviews with a young man from the Philippines who climbed the career ladder very quickly since he is now the president of a leading agricultural college in his country.

My eventual involvement in scientific agriculture had been largely dictated by circumstances rather than personal choice. Despite my farming background and my boyhood in a rural setting, my first choice was engineering because of my love for machines. However, my parents would have had to spend more money and the University of the Philippines at Los Baños was an attraction to agriculture because of its prestige. Graduate work was not in my plans at the start, but there remained my interest to discover new ideas, new things, and research was a good medium to execute ideas. In short, research led to higher education and higher education led to more research, better discoveries and contacts with other scientists. Needless to say, recognition of my work by institutions, colleagues, and farmers drove me harder into research. However, I found out that I could not stay forever in one place without a break. Foreign assignment lured me as a way of discovering new things, learning new ideas, making new contacts, and, of course, bringing satisfaction not only to me but also to the members of my family. The economic reward cannot be ignored.

Thanks to contacts at Texas A&M University where he had specialized in soil and crop science, he easily found a job in a U.S.AID program in Haiti for two years before returning to the Philippines. Many scientists from the DCs are haunted by the desire to spend some time working outside of their home country. During my missions many of them asked me whether IFS or some other international organization might be able to use their services. One major reason is that salaries paid to national research scientists in many DC countries, especially in Africa, are inordinately low.

Remuneration

It is very difficult to obtain a clear view of the problem of wages for research scientists and, for a multitude of reasons, even more difficult to make comparisons between countries: different standards of living, convertibility of local currency, the effects of a black market, etc. I tried to avoid the problem by asking our respondents to compare their wages with the minimum national wage. I then tried to match the answers to this question with the question on whether the salary they received was "adequate" or "inadequate." (These were the only two choices.) Of course, there was no question of making comparisons with salaries paid in the developed countries.

Before trying to interpret the results of table 3.2, we had to look at the DCs that do not have a minimum guaranteed salary, countries that base salaries on the supply and demand of the labor market. Since in the DCs supply far exceeds demand, salaries are very low. In French-speaking Africa, for instance, the minimum monthly salary is somewhere around the equivalent of U.S. $100. Most of the scientists in our study population earned between 1 and 5 times this amount, like most of the other scientists in the country, since close to two-thirds of them (64.92%) received between 1 and 5 times the minimum wage. Table 3.3 shows the breakdown of salaries.

Table 3.2. Wages Paid to Research Scientists: Comparison with
Minimum Wages in DCs and Relative Satisfaction

Multiple of Minimum Wage	Adequate Satisfaction		Inadequate Satisfaction		Total	
1–5	82	27.5%	216	72.5%	298	64.9%
6–10	69	53.1%	61	46.9%	130	28.3%
11 and more	18	62.5%	13	37.5%	31	6.8%
Total	169	36.8%	290	63.2%	459	100.0%

Table 3.3. Salaries Paid to Research Scientists

Multiple of Minimum Wage	No. of Scientists	Percentage
1	12	2.4%
2	61	13.2%
3	80	17.5%
4	75	16.3%
5	71	15.5%

The minimum salary varies greatly within a broad spectrum. The
lowest salaries are paid in India and the highest in certain countries of
Latin America and in Nigeria where an oil boom led to such inflation
that salaries were increased every week or fortnight in an unsuccessful
attempt to keep up with increases in the cost of living.

At the top of the scale, we see that there are very few (6.76%)
scientists who earn as much as 11 times the minimum wage. The
highest wage was obtained by the dean of a faculty in Nigeria who
earned 99 times the minimum wage, followed by a scientist in India (80
times) and another in Malaysia (40 times). I was not able to check these
figures, and I quote them for their indicative value only, with due
caution and reservation. Most scientists in the "over 11 times" group
earn between 15 and 20 times the minimum wage and reside in 19
countries located on the three continents, but we have not noticed a
significant concentration in any given country.

It is not surprising that the higher the number of times a scientist's
salary was over the minimum wage the more adequate he felt it was.
We noted that over two-thirds (72.48%) of the researchers who earned

Table 3.4. Nature of Second Job and Number of Hours Spent

Type of Side Job	Hrs. Spent Weekly on Side Job									
	1–5		6–10		11–20		21 and More		Total	
Consultancy	16	26.2%	22	36.1%	18	29.5%	5	8.2%	61	38.6%
Teaching	20	40.0%	14	28.0%	11	22.0%	5	10.0%	50	31.6%
Agriculture	2	10.5%	4	21.1%	8	42.1%	5	26.3%	19	12.0%
Commerce	1	5.6%	8	44.4%	6	33.3%	3	16.7%	18	11.4%
Research	0	0.0%	3	—	2	—	1	—	6	3.8%
Translations	0	0.0%	2	—	0	0.0%	0	0.0%	2	1.3%
Other	0	0.0%	2	—	0	0.0%	0	0.0%	2	1.3%
Total	39	24.7%	55	34.8%	45	28.5%	19	12%	158	100%

between 1 and 5 times the minimum wage felt that their salary was inadequate, a feeling that was shared by close to two-thirds (63.18%) of the whole study population, without reference to the multiple of the minimum wage. This explains why many of them supplement their income by working overtime on side jobs that include anything from consultant in their specialty to teacher to taxi driver.

Close to one-third of the IFS grantees said they had side jobs. This percentage may seem high, but we think that the number is actually higher because for most of them it was illegal to have a second job, especially those holding a civil service position. None of the scientists in India confessed to having a side job, and many of them in India and in other countries such as Morocco actually insisted that it was quite impossible to have a side job since it was illegal.

Anyone who has spent time with DC scientists quickly realizes that a second (or even third) job and income are vital. Some of the scientists went so far as to take us to their second place of work so that they could continue discussing their IFS-funded project with us.

The question on the nature of the work is still open, but the types of work easily fit into four main categories (see table 3.4) that account for over 90% of the total: consultancy (38.61%), teaching (31.65%), agriculture (12.02%), and commerce (11.39%). Consultancy work is nearly always related to the fields developed in the research activity. In aquaculture it is quite normal for a scientist working on the optimal postlarvae production conditions for king prawns (*Macrobrachium rosenbergii*) to offer his services to a local shrimp breeding enterprise in which he may eventually buy shares. Two- thirds of the scientists who also work as consultants spend a notable 6 to 20 hours a week on their outside job.

Teaching is less time consuming since over two-thirds of the scientists who supplement their normal jobs by teaching spend only between 1 and 10 hours a week on it. Teaching can mean working overtime at one's normal place of work or, as is more often the case, at an outside institution.

Agriculture can involve anything from working on a coffee plantation to raising layer hens. A Ghanaian scientist in Kumasi one day showed us the hens he farmed in his backyard. The eggs were sold at the local market and permitted him to double his income. Scientists working on commerce usually were lending a hand in a family enterprise.

Because salaries are unacceptably low, the question of additional income was repeated as a leitmotiv in the comments the scientists added to the questionnaire, as we can see from the following examples:

"The economic situation in my country and in many other DCs over the last ten years has made it very difficult to work as a researcher. Salaries are too low. At present [around 1985] the net monthly wage for a scientist in Ghana is 2,500 Cedis [about U.S. $50]. This wage is so low that the scientist spends much of his time looking for another source of income."

"The very poor salaries paid to university professors in Peru forces them to look for other sources of income to support their families, and therefore they cannot spend a proper amount of time on their research. An unskilled worker gets a quarter of the salary paid to a senior professor with 17 years of service behind him."

"Salaries paid to research scientists in Indonesia are so low that most of my colleagues have left the university to work for private firms."

"The main constraint is the ridiculously low salaries that are paid to scentists in Sri Lanka. A bank employee with only primary schooling earns a higher salary, without even counting the other advantages he can enjoy. We scientists spend our holidays, our evenings, and our weekends working for private companies and teaching as visiting lecturers just to be able to buy enough food to feed ourselves."

Outside jobs were of equal concern to scientists who studied abroad and to scientists who studied exclusively in their home country, but it seemed clear that the longer a scientist had been abroad the less satisfied he was with his salary. We were able to confirm that the longer a scientist studied abroad the greater his qualifications and the higher his rank in the hierarchy, which also meant a relatively higher salary. We can also postulate that for scientists who spent many years abroad where they enjoyed certain economic benefits and the wages paid in the developed countries it was difficult to accept the lower standard of living that awaited them upon their return home.

Choosing the Research Topic

We saw that the decision to enter research and the choice of disciplines can be influenced by the possibility of obtaining a scholarship at a certain time in one's school life. This factor seems to be so decisive that some authors go as far as to claim that "the fact that a scientist is engaged in a particular scientific specialisation does not necessarily mean that he is very interested in it" (Eisemon 1979, p. 515). Does this also apply to research topics? Various criteria have been considered decisive.

In a recent study Zuckerman (1978) concluded that there were two main criteria, namely, the evaluation and perception of the scientific importance of a problem and the feasibility of solving it. Medawar (1967, p. 7) also underscored the importance of only choosing research topics that had perceivable solutions when he said, "If politics is the art of the possible, research is surely the art of the soluble."

Scientific literature is often inspiring for the scientist who is trying to identify a problem, as a Cuban IFS grantee explained. "When I left the university in January 1974," he said, "I was hired at the Botany Institute. I was put in charge of the Plant Eco-physiology Department in 1980. When I first started working at the institute, I was asked to work on mycorrhiza. First I said no, claiming that I didn't know anything about the subject. The Director of the Institute then gave me an article by Barbara Mosse that was published in 1972 on the effects of various strains of mycorrhizal fungi on the growth of *Paspalum notatum*, a fodder grass commonly found in Cuba."

A research topic might also be chosen because there is a fair chance of having the results published in a specialized journal. This idea was set out in an Indian research publication cited above in an interview with an Indian physicist who said, "We draw up projects by looking through papers in the 'Physical Review,' the 'Journal of Physics,' etc. to find out what types of things are being done, because if you don't do that, you don't get a job here. You have to publish in those journals, so you must do something which they are doing." (Shiva and Bandyopadhyay 1980, p. 583).

The choice of a topic may also depend on the equipment available for research. A grantee from Costa Rica who was in Belgium working on the chemistry of synthetic products mainly applied to antibiotics had to stop his research when he returned to his country because of lack of proper infrastructure. He decided to switch to the chemistry of natural products because someone else from the department of chemistry had just returned from a year's specialization in this field in the United States and the equipment required to begin the research was available.

Another factor that plays a considerable, even a vital, role in choosing a subject for research is funding; the researcher's need to accept some concessions is discussed in chapter 4.

We have just seen that the choice of a research subject can be influenced by a series of factors, some of which are external to the science involved. In fact, most often several factors intervene simultaneously in this choice: "Decisions made by scientists regarding problem choice emerge from a complex process of negotiation within themselves and with other scientists, administrators, and clients" (Busch and Lacy 1983, p. 44).

In an attempt to determine the relative importance of the different factors that may have played a part in the choice of research subject of IFS grantees, we have adapted Busch's list of criteria that has been tested in the United States and in Sudan (Busch and Lacy 1983, p. 45) to the requirements of our study by both eliminating and adding a few factors. Our basis was the mean figures obtained from a system based on a scale of five numbers (running from 1, which signified "essential," to 5, which signified "not important at all"). We established a classification of the 20 choice criteria offered to the researchers, as can be seen in table 3.5.

The leading criterion, "importance to society," takes us back to the criterion that was in second position in the list on choosing research as a profession, namely, "social utility." Here again we see the scientist's need to justify himself to society. When we asked the grantees what this concept meant to them, they indicated that social utility was more or less the capacity of research to solve the economic and social problems facing their country. This criterion is rather close to other criteria in the list, e.g., likelihood of clear empirical results (rated ninth) and to a lesser extent, potential marketability of the final product (rated twelfth).

The fact that the top four criteria refer to a fairly heterogeneous assortment of concepts is probably significant and indeed tends to confirm the hypothesis that the choice of a research subject depends more on a series of factors than on any one single factor.

It is also interesting to note that the position of the first six criteria on our list corresponds (although in a different order as concerns the first three and excepting the fifth criterion, which is different on the two

Table 3.5. Research Problem Choice: Relative Importance of Various Criteria

Rank DC	Criterion	Mean Score[a]	Busch class 'n USA[b]
1	Importance to society	1.76	2
2	Potential creation of new methods, useful materials, and devices	1.88	5
3	Enjoy doing this kind of research	2.06	1
4	Scientific curiosity	2.23	4
5	IFS priority area	2.28	—
6	Publication probability	2.34	6
7	Availability of research facilities	2.40	3
8	Potential contribution to scientific theory	2.45	12
9	Likelihood of clear empirical results	2.65	8
10	Access to external funding	2.67	(9)[c]
11	Currently a "hot topic"	2.71	15
12	Potential marketability of final product	2.71	17
13	Priorities of the research organization	2.71	11
14	Length of time needed to complete project	2.82	16
15	Credibility of other investigators doing similar research	3.07	14
16	Access to funding from your institution	3.21	(9)[c]
17	Feedback from extension personnel	3.30	20
18	Colleagues' approval	3.33	18
19	Subject of your thesis	3.49	—
20	Demands raised by clientele	3.59	13

[a]The mean is based on a 5 number scale: 1 = essential, 2 = very important, 3 = moderately important, 4 = not very important, 5 = not important at all.
[b]Busch classification was based on a sample of 1,431 research scientists and used a series of 21 criteria.
[c](9) = Funding only.

lists) to what Busch and Lacy (1983) established in the United States when working on a sample of 1,431 American researchers employed in the agricultural sciences. From a broader comparison that includes the list as a whole, we see that, with but a few exceptions, the two lists are relatively similar. This leads us to believe that the IFS grantees in our population have more or less adopted the same reference systems as concerns criteria used for choosing research subjects as American researchers working in comparable fields. Furthermore, when assessing whether studying abroad affected the relative importance assigned to the first ten criteria, we found that there was no really significant difference in ratings by students, whether they had studied—for varying periods of time—abroad or even had not studied abroad at all. Table A5 in Appendix C, for instance, shows that the number of years spent abroad does not have a significant effect on the importance assigned to the criterion "potential contribution to scientific theory" used in choosing a research subject.

The DC scientists in our population assign more importance than their American colleagues to criteria such as "potential creation of new methods, useful materials and devices," and "potential contribution to scientific theory," which are more characteristic of basic research. On the other hand, the fact that the criterion "demands raised by clientele" is at the bottom of our list no doubt reflects the marginal position of science in the DCs and supports the theory that research scientists and scientific institutions are kept out of the production lines because of the absence of a "demand-pull" the economic system could have on the local know-how production system.

It is also probably significant that the importance of "the availability of research facilities" is underestimated by the scientists, who rate it seventh on the list of criteria used in choosing the research subject, and yet second on the list of constraints to progress in their research work. Perhaps there is an effort to mask reality when choosing a research subject so as to gloss over the objective impossibility of carrying out a given piece of research even if there is a risk of having to face harsh reality later. It also seems significant that the "IFS priority area" comes fifth and "access to funding from your institution," sixteenth. The proportion of external financial aid represents an ever larger percentage of research budgets in the DCs and, *pari passu*, influences the choice of research subjects. Chapter 4 looks at this subject in greater detail.

In an endeavor to determine the relative influence of various people on the choice of the research subject, we used the same method as for choosing the research subject. We prepared a series of 12 people or groups of people on a list that we submitted to the grantees. The results (table A6, Appendix C) indicated that the people in the series had little or no influence on the grantees and that there was little variation in the answers from whch we calculated the means. The person who had the biggest influence on the scientist's choice of research subject was his immediate superior whose mean score was 3.31, calculated on the basis of the following ratings: 1 = essential, 15%; 2 = very important, 20%; 3 = moderately important, 17%; 4 = not very important, 15%; 5 = not important at all, 33%.

Actually, many respondents added their own names to the list,

Table 3.6. Relation of Thesis Subject to IFS-Funded Project as a
Function of Country of Doctoral Studies

Country of Doctoral Studies	Direct Relation		Little or No Relation		Total
IC	107	50.5%	105	49.5%	212
Asia	29	87.9%	4	12.1%	33
Africa	24	88.9%	3	11.1%	27
Latin America	1	—	3	—	4
Total	161	58.3%	115	41.7%	276

indicating that they were the person who had had the greatest influence on the choice of research subjects. The thesis supervisor and the subject of the thesis were not reported to have much effect on the choice of the research subject since the mean scores were 3.72 and 3.49, respectively. Yet I was able to confirm that in close to 60% of the cases the subject of the thesis was directly related to the work that the scientist would be researching in his home country after completing his thesis.

The percentage varies considerably as a function of place of study, i.e., in an industrialized country (IC) or DC. For half the students who studied for their doctorate in an IC, as table 3.6 shows, there was little or no connection between the IFS-funded research and the subject of the thesis, while close to 9 out of 10 doctorates obtained in Asia (87.88%) and in Africa (88.89%) were directly related to the IFS-funded research project.

There are too few doctorates from Latin America to be able to calculate any significant percentages. These results seem very important to us since a large number of scientists who studied in ICs had to change research subjects when returning to their home countries. There were many reasons (table 3.7), but the main one was related to matching research work with national needs.

A researcher who had worked on the problem of nutrition linked to obesity in the United States quite obviously had to change subjects upon his return to Thailand; he decided to work on controlling the thiamine (vitamin B_1) deficiency caused by consuming too much tea and tannin.

The initial reflex of an African researcher who had studied *in vitro* cultivation of endives in France was to continue his work when he returned to Congo, especially since he saw that endives grew well there. After further thought he decided to study the plant growth

Table 3.7. Reasons for Changing Research Subjects

Classi-fication	Reason	No. of Scientists	Percentage
1	To work on problems of local importance related to my country's needs	46	38.3
2	To work in more applied research	17	14.2
3	Lack of equipment	17	14.2
4	To work in a new field	12	10.0
5	To expand my knowledge to other fields	7	5.8
6	To participate in a conference	6	5.0
7	Possibility of obtaining funding	5	4.2
8	Other	10	8.3
	Total	120	100.0

physiology of *Gnetum africanum*, a local leafy vegetable that was disappearing. In 1985 he obtained his Doctorat d'Etat from the Université d'Orléans after defending a thesis on this subject.

Why scientists change subjects is an open question. I did not have much difficulty in grouping the answers, except as concerned the fourth reply, "to work in a new field." The scientists in this group, i.e., 10% of all those who changed research subjects, gave rather varied answers that were not always easy to interpret. In some cases it merely meant changing subjects; in other cases it meant initiating work in a field that was new to the country. Part of the answers probably could have been grouped with the answers at the top of the classifications list, but for reasons of precision I decided to isolate this group. If we recognize that the two reasons that tied for second position are closely correlated to the first reason, we can conclude that close to two-thirds of the researchers changed subjects in order to adapt to local conditions, i.e., for reasons of relevance or lack of resources.

Returning to the subject of the thesis, we see that the more recent the date of the doctorate the more direct the relationship is between the subject of the thesis and the current research work (table 3.8). These results were as expected and could be explained by essentially three reasons.

First, the percentage of doctorates awarded in the DCs is increasing (chapter 2, table 2.6), and we have just seen that some 90% of the subjects of doctoral theses completed in the DCs were directly related to the scientists' current research. Second, increasing attention is being given in the developed countries to the relevance of the thesis topic;

Table 3.8. Relation of Thesis Subject to IFS-Funded Research as a
Function of Year Doctorate Obtained

Year of Doctorate	Direct Relation		Little or No Relation		Total
1960–69	18	48.7%	19	51.3%	37
1970–74	27	42.9%	36	57.1%	63
1975–79	47	54.0%	40	46.0%	87
1980–85	66	76.7%	20	23.3%	86
Total	158	57.9%	115	42.1%	273

Table 3.9. Relation of Thesis Subject to IFS-Funded Research as a
Function of Discipline

Discipline	Direct Relation		Little or No Relation		Total
Aquaculture	24	55.8%	19	44.2%	43
Animal production	28	71.8%	11	28.2%	39
Crop science	59	71.1%	24	28.9%	83
Forestry	12	60.0%	8	40.0%	20
Food sciences	17	56.7%	13	43.3%	30
Natural products	37	50.7%	36	49.3%	73
Rural technology	4	40.0%	6	60.0%	10
Total	181	60.7%	117	39.3%	298

and even when the research for the thesis is done completely in a
developed country, special efforts are made to allow the student/
researcher to work in experimental conditions that are as similar as
possible to those prevailing in his home country. In crop science, for
instance, we closely monitored various research activities leading to a
thesis in which the student had imported plant material from his home
country and used it in greenhouse or growth chamber conditions
where he could control the temperature, light, and hygrometry. Third,
a research scientist, as he progresses in his career, is justified in changing research subjects; it would be abnormal for him not to do so.

The relation of the thesis topic and the IFS-funded research varies
according to subject, as we can see in table 3.9. The closest links are
found in fields like agronomic research in the broad sense of the term

and in rural development, i.e., animal production, crop science, and forestry.

Many scientists now working on aquaculture, a relatively recent research area requiring a multidisciplinary approach, were trained in a related discipline such as oceanography, hydrology, marine biology, or fisheries. After completing their studies, not without difficulty, they switched to aquaculture. There are a few universities highly reputed in the field of aquaculture, such as Auburn in the United States and Stirling in Great Britain. Other efforts to create regional training centers to work with research centers on aquaculture in the DCs, e.g., the Philippines, China, Nigeria, and Brazil, are very recent.

Rural technology is the field where we find the weakest link between the thesis topic and the IFS-funded research. We have the example of the African researcher who used to work on theoretical physics applied to nuclear energy in France. Upon his return to Senegal, he had to start working on the solar energy requirements of a program to design solar dryers for fish and other foodstuffs. The last time we met was in 1985 in Dakar. Unfortunately, he had stopped all research activities because his teaching obligations did not leave him enough time. He wanted to obtain a Doctorat d'Etat from the Université de Nice in France in order to become a full-fledged professor but felt that the results of research on solar energy would not generate enough data for a doctoral thesis. Thus, he went back to his original specialization, nuclear physics, and prepared a thesis on the radiation of uranium and radium.

The last example reveals the importance and the influence of the environment in which the researcher works on the research methods and the choice of research subjects. Research scientists devote varying amounts of time to research. Much depends on the objectives of their home institute as we will see by the following.

Institutional Contexts

The institutes where our grantees work can be grouped as shown in table 3.10. We made a distinction between agricultural colleges and "general" universities. The former usually have research programs that are defined in close conjunction with development targets. They also usually have experimental farms and test plots and, in certain cases, their own extension services. The main ones include the Tamil Nadu Agricultural University at Coimbatore, India, the College of Agriculture of the University of the Philippines at Los Baños (UPLB),

Table 3.10. Institutions Where Research Scientists Work

Type of Institute	No. of Scientists	Percentage
University (general)	261	53.4%
National research institute	109	22.3%
Agricultural college	55	11.2%
Research institute in a university	33	6.7%
Ministry	12	2.5%
Development agency	7	1.5%
Regional research institution	5	1.0%
Private institution	2	0.4%
Other	5	1.0%
Total	489	100.0%

Philippines, and the Hassan II Institute of Agronomy and Veterinary Sciences at Rabat, Morocco.

We have established two subgroups for the research institutes: national research institutes and university research institutes. Although the latter generally have their own statutes that guarantee their independence from the host university, they usually maintain preferred relations with the university that entail admitting postgraduate students to their research programs and teaching. One of the most dynamic examples is the Institute of Chemistry at the University of Karachi in Pakistan.

Research in small countries that do not have national research institutions can be conducted under the auspices of a ministry or a development agency or service. But it must be kept in mind that the main function of the latter is development or extension, not research. Regional research institutions are new to the DCs. Quite a few already exist in Southeast Asia, e.g., the Southeast Asian Fisheries Development Center (SEAFDEC), which has two main centers, one in the Philippines and one in Thailand. Since they are generally properly funded, IFS has made no special effort to consider candidates from among the young researchers working there.

Nearly all the scientists in our study population (97.5%) work for government institutions, and about two-thirds (71.37%) of this group work in a university establishment. Let us see what effects this can have on the time devoted to teaching and thus, indirectly, on their research (table 3.11).

The first conclusion that can be drawn from table 3.11 is that more than half the researchers (55%) devote a hefty 20% to 60% of their time

Table 3.11. Percentage of Time Spent Teaching According to Host Institution

% of Time Spent Teaching	University		Agricultural University		Research Institute within a University		National Research Institute		Total	
0	11	4.2%	1	1.8%	10	30.3%	57	52.8%	79	17.4%
1–20	28	10.8%	8	14.8%	12	36.4%	45	41.7%	93	20.4%
21–40	94	36.2%	27	50.0%	9	27.3%	4	3.7%	134	29.4%
41–60	97	37.3%	15	27.8%	2	6.0%	1	0.9%	115	25.3%
61–80	27	10.4%	3	5.6%	0	0%	1	0.9%	31	6.8%
81–100	3	1.1%	0	0%	0	0%	0	0%	3	0.7%
Total	260	100.0%	54	100.0%	33	100.0%	108	100.0%	455	100.0%

to teaching activities. Obviously, the heaviest teaching load is shouldered by researchers working in universities where close to half the researchers spend more than 40% of their time teaching. The time devoted to teaching seems to be relatively less in agricultural universities than in other universities. Researchers in research institutes located on university grounds sometimes spend a nonnegligible part of their time teaching, although 30% of them do not teach at all. It is important to observe that there is no hard-and-fast barrier separating the universities and the research centers since more than 40% of the researchers in the research centers devote between 1% and 20% of their time to teaching.

The percentage of time spent on teaching obviously also depends on the researcher's functions within his home institution. This largely explains why 4% of the scientists in the universities do not teach at all, viz., most of them hold high-level posts involving administrative duties that leave little or no time for teaching or research.

Table 3.12 shows the 10 main functions of IFS grantees in their universities and research institutes. Classifications have been worked out using equivalences for titles and thus are based partly on approximations.

The first observation that we can make is that most of the scientists (90%) still have functions that should allow them to conduct research. Percentagewise the most widespread function in the university is senior lecturer (27.1%); and in the research institutes, the main function is research officer (11.2%).

Checking the effects of years spent abroad on the functions of the

Table 3.12. Present Employment as Related to Years Spent Abroad

| Present Employment | Years Abroad | | | | | |
	0	1-2	3-4	5-9	10-20	Total
University						
Lecturer	21	12	24	21	4	82
	35.6%	12.4%	21.3%	15.6%	16.6%	19.2%
Sr. lecturer	13	21	33	43	6	116
	22.0%	21.6%	29.2%	31.9%	25.0%	27.1%
Professor	6	15	25	32	6	84
	10.2%	15.5%	22.1%	23.7%	25.0%	19.6%
Dean	0	0	1	2	2	5
	—	—	.9%	1.5%	8.3%	1.2%
Vice chancellor	0	0	1	3	1	5
(Dep. vice chanc.)	—	—	0.9%	2.2%	4.2%	1.2%
Research Institute						
Research assist.	4	7	1	0	0	12
	6.8%	7.2%	0.9%	—	—	2.8%
Research officer	9	18	12	8	1	48
	15.2%	18.6%	10.6%	5.9%	4.2%	11.2%
Sr. research officer	1	10	6	4	1	22
	1.7%	10.3%	5.3%	3.0%	4.2%	5.1%
Lab. supervisor	5	11	5	11	0	32
	8.5%	11.3%	4.4%	8.1%	—	7.1%
Director of	0	3	5	11	3	22
institute	—	3.1%	4.4%	8.1%	12.5%	5.1%
Total	59	97	113	135	24	428
	100.0%	100.0%	100.0%	100.0%	100.0%	100.0%

scientists, we saw that in the university the effects were greatest on professors. Hence, the person who has spent between 10 and 20 years abroad has two and one-half times more chance of becoming a professor than the person who completed all of his studies in his home country. The influence is even clearer for the position of dean and vice-chancellor. For the latter two functions, however, our results come from small samples; and we should avoid exaggerated generalizations.

For the research institutes, the post of director seems to be the one that is most strongly affected by stays of varying duration abroad. All of the 22 IFS grantees who became research institute directors had spent some time studying abroad. A scientist who studied abroad for 10 to 20 years had four times more chance of becoming the director of a research institute than the scientist who studied abroad for 1 or 2 years.

Here again we have to be careful in interpreting these results because a post assignment is the result of a whole gamut of interacting factors. We saw previously that obtaining a doctoral degree could be related to the number of years spent abroad (chapter 2, table 2.7). And the number of years spent abroad can also be correlated with age. It is in the 40 to 49 year age group that we find the highest number of scientists who spent at least 10 years abroad (table A7, Appendix C).

Chapter Four

Practicing Research

Time Devoted to Research

In the preceding chapter we saw that the time devoted to teaching depended on the nature of the scientist's home institution. We can say the same for the time devoted to research, which, moreover, is inversely proportionate to the time spent on other functions in the institution such as teaching and administration. It also depends on the type of job the researcher or teacher/researcher has in the institution as table 4.1 clearly shows.

As one might expect, it is the young scientists early in their careers who devote the most time to research, e.g., research assistants, who spend more than 60% of their time on research activities. In the universities, professors form the category that devotes the most time to research, although only about 12% spend more than 60% of their time on it. We can thus assume that professors turn over part of their teaching obligations to senior lecturers and even more so to lecturers. It is interesting that the vice-chancellors and their deputies reserve at least a small amount of time for research. The directors of certain research institutes seem to participate very actively in research since 30% of our respondents told us that it consumed 30% to 60% of their time, and 6% reported spending more than 60%. A partial explanation may be that some research institutes in the DCs are small and short of scientists. The directors of such institutes devote considerable time to research partly because they are not bogged down by their administrative duties and partly to compensate for the shortage of scientists. Further, most of the scientists in our sample are working in countries where research institutes are quickly developing and growing in numbers. The result has been that they rapidly climbed up the career ladder to reach posts of very great responsibility, in some cases without having had enough time to confirm their research capabilities.

We have often seen that the average career of a DC scientist is shorter than that of a scientist in an IC. Let us look at two examples. The first one concerns a young West African research scientist who, after lengthy studies in France, returned to his home country and within seven years became department head, dean of a university faculty, and

Table 4.1. Percentage of Time Spent on Research in Terms of Present Employment

Present Employment	Percentage of Time for Research						Total
	0–30		31–60		61–99		
University							
Lecturer	32	39.0	48	58.6	2	2.4	82
Sr. lecturer	53	45.7	57	49.1	6	5.2	116
Professor	33	38.8	42	49.4	10	11.8	85
Dean	4	80.0	1	20.0	0	0.0	5
Vice-chancellor (Dep'y vice-ch'r)	4	80.0	1	20.0	0	0.0	5
Research institute							
Research assistant	0	0.0	3	25.0	9	75.0	12
Research officer	6	12.5	18	37.5	24	50.0	48
Sr. research officer	2	8.7	9	39.1	12	52.2	23
Lab. supervisor	4	12.1	18	54.6	11	33.3	33
Director of institute	11	64.7	5	29.4	1	5.9	17
Total	149	35.0	202	47.4	75	17.6	426

then vice-chancellor. The second concerns a young East African scientist who, after graduating with an M.S. from an American university, returned home and was almost immediately put in charge of a nation-wide plant protection service that meant devoting much of her time to administration. One year later she was awarded a scholarship to study for a Ph.D. in the United States. When she returned home, she was appointed head of research for her whole institute.

It is both symptomatic and paradoxical that the research scientists who spend 10 and more years abroad learning research are the ones who spend the least time on research after returning to their home country (see table A8 in Appendix C) and the most time on administration (table 4.2).

Thus, in the category of researchers who devote more than 40% of their time to administration, there are proportionately five times more who spent 10 or more years studying abroad than those who studied solely in their home country. Between these two extremes runs an almost exponential progression of scientists working in administration that is a function of the number of years spent abroad.

It should be noted that for close to 60% of the scientists in our sample

Table 4.2. Time Spent on Administration by Number of Years Spent Studying Abroad

No. of Years Abroad	Percentage of Time on Administration						Total
	0–10		11–40		41–99		
0	52	71.2%	18	24.7%	3	4.1%	73
1–2	69	61.1%	34	30.1%	10	8.8%	113
3–4	64	54.7%	41	35.0%	12	10.3%	117
5–9	82	54.3%	48	31.8%	21	13.9%	151
10–20	11	40.8%	9	33.3%	7	25.9%	27
Total	278	57.8%	150	31.2%	53	11.0%	481

administrative duties consume less than 10% of their time; and for slightly over 10% of the scientists, administrative duties consume over 40% of their time.

Here again, we realize that greater precision would require a multifactorial analysis. Yet it is interesting to observe that training abroad seems to confer eligibility for a position of responsibility for which the scientist has no exceptional qualification. Actually, the purpose of going abroad is to learn research, not research administration; these are two very different activities that require very different capabilities.

We have seen previously that among the criteria used to choose research topics the scientists in our population favored criteria that were more or less directly connected to, or seemed to be characteristic of, basic research. It is practically impossible to find functional definitions that rigorously distinguish "applied" from "basic" research. Pasteur was no doubt right in stressing that "there is no single category of science which we can call 'applied science': there is science and there are applications of science, the two being linked as a fruit is to the tree that bears it" (cited by Salomon 1970, p. 141). Furthermore, science considered as basic in one context may be considered as applied in another. The IFS has a small number of detractors who say its research program leans too heavily toward applied science, while others say its (over)emphasis on basic science is detrimental to a development focus. The latter point of view is shared by a handful of Dutch and Belgian administrators. Belgium, partly for this reason, decided to stop financing IFS activities, just a few months before the King Baudouin Foundation awarded IFS the King Baudouin Prize for development!

It seems significant that a not negligible number of researchers found it difficult to distinguish between the time spent on "applied"

Table 4.3. Apportionment of Time among Teaching, Basic Research, Applied Research, and Development (average percentages)

	In department or research group		In IFS-supported research program	
	Present	Ideal	Present	Ideal
Teaching	38.1	29.1	25.3	20.9
Basic research	17.4	21.7	21.8	23.2
Applied research	29.9	32.3	37.5	37.7
Development	12.6	16.9	12.5	16.8

and on "basic" science. In an attempt to assess the overall orientation of their research and to help them with this breakdown, we suggested the following definitions, adapted from those used by the National Science Foundation (Kidd 1959):

Basic research stresses increases of knowledge in science, with "the primary aim of the investigator being fuller knowledge or understanding of the subject under study, rather than a practical application thereof."

Applied research is directed toward practical application of knowledge. It covers "research projects that represent investigations directed to discovery of new scientific knowledge and that have specified commercial objectives with respect of either products or processes."

Development may be summarized as "the systematic use of scientific knowledge directed toward the production of useful materials, devices, systems, or methods, including design and development of prototypes and processes."

We asked the grantees to indicate the apportionment of their time (table 4.3) between teaching, basic research, applied research, and development activities, using the definitions set out above. Further, to find out how satisfied they were, we asked them to characterize the ideal time apportionment. In an effort to work out a comparison with colleagues outside of IFS, we asked about the distribution of time within their department or research group. It was obviously not possible to check the results since they represented the personal assessment of each scientist we consulted.

The first clear observation is that IFS grantees seem to devote more time to research (an average of 60%) than their colleagues working in the same department or research group and thus spend less time teaching. On the whole they felt that they devoted more time to applied

Table 4.4. Breakdown of Time Devoted to Basic Research, Applied
Research, and Development: Comparison of IFS and American
Research Scientists (Average Percentages)

	IFS	American
Basic research	30	30
Applied research	52	55
Development	17	13

research (37.5%) than to basic research (21.8%) or to development
(12.5%). Were they free to choose, they would spend less time on
teaching, more time on basic research, and much more time on development. They seem generally satisfied with the amount of time they
had for applied research. For purposes of comparison with the results
observed by Busch and Lacy (1983, p. 66) in the United States, we
discounted the time spent on teaching and recalculated the percentage
of time spent on research and development on a basis of 100. The
comparison is again quite instructive as table 4.4 shows us.

Thus, the scientists in our group and American scientists working in
comparable fields spend exactly the same percentage of their time on
basic research. The time our group spends on applied research is
slightly less and the time on development, slightly more, although the
differences are not significant. Here again DC and American researchers
seem to feel almost identically about the research they practice.

These results are means and thus conceal interdisciplinary differences. Chemistry and microbiology research scientists claim to
spend at least 50% of their time on basic research. Conversely, scientists
researching crop science and animal production devote 15% to 20% of
their time to basic research. Busch found the same time distribution
among the American research scientists.

To conclude, let us consider where the scientists worked during the
12 months immediately before the survey:

in the laboratory	30.3%
in the office	22.8%
in the field	20.0%
in the library	14.9%
in animal and plant experimental areas	5.3%
elsewhere	3.5%
in the computing facilities	3.2%

Table 4.5. Expenses for R&D Expressed as a Percentage of GNP

	1970	1975	1980
ICs[a]	2.36	2.25	2.24
DCs[b]	0.30	0.36	0.43
Africa	0.33	0.35	9.36
North America	2.47	2.13	2.18
Latin America	0.32	0.51	0.53
Arab states	0.31	0.23	0.27
Asia	1.02	1.08	1.18
Europe	1.70	1.78	1.79
Oceania	1.10	1.04	1.11
USSR	4.04	4.79	4.67

[a] All the European countries, USSR, United States, Canada, Japan, Israel, Australia, and New Zealand.
[b] All the other countries in the world.

This is the distribution of time, expressed as percentages, devoted to research for all disciplines combined. A finer analysis would undoubtedly bring out more details on differences between the disciplines, probably showing that differences could be related to level of education or position within the institute.

Funding Research

In chapter 3 we hastily observed that access to funds was a criterion that affected the choice of research topic and that its importance was greatest when funding was obtained from IFS, that it was less when it came from an institution other than the grantee's home institute, and that it was least when it came from the budget of the grantee's institute.

After this abridged picture of R&D funding in DCs, let us look at the size of the budgets available to grantees before and after receiving the IFS grant, the relative importance of different sources of research funds, and the effects on the choice of research topic and method.

The 1985 UNESCO statistics (table 4.5) show that R&D outlay in the DCs was estimated on the average to be 0.30% of the GNP in 1970 and rose to 0.43% in 1980. During this same period of time, outlay for R&D

in DCs as compared to worldwide outlay rose from 2.3% to 6%. Globally, R&D has been allocated greater financing in the DCs, but the benefits have not been the same everywhere. In Latin America and Asia, R&D funds doubled between 1970 and 1980, while in Africa the increase was very slight. A more detailed study shows that a goodly part of R&D investments in the DCs was consumed in countries such as India and Brazil.

The last decade has been marked by considerable change. At the beginning of the 1970s, efforts were made systematically at both the national and the international levels to promote and support R&D activities in the DCs. The efforts were expressed through the creation of new organizations and an increase in the number of donors and their overall financial contributions.

External funding agencies have been paying for increasing parts of research budgets in the DCs. In the field of agricultural research in the DCs as a whole, the figure has reached some 40% (Oram 1985). In some African countries such as Mali, Mozambique, Senegal, Lesotho, Swaziland, and Zambia, the contribution has been as high as 70% and even higher. Here again the difference is very great since foreign aid only accounts for less than 15% of the national R&D budgets in Cameroon and Sudan. In some countries the number of donors involved in research financing is so great that it is practically impossible to determine the share that comes from the national budget. In any case, this was the conclusion drawn in a recent report by ISNAR (1983) on research on agriculture and livestock in Burkina Faso. The report stated that each year this little country received no less than 340 visits from foreign governmental, multilateral, and international agencies for development-oriented research.

This erratic growth of foreign aid and the absence of coordination cause serious problems, e.g., the capacity to absorb this aid, the excessive numbers of task forces and individual visits for a variety of aid programs and some small projects, and the impossibility of integrating aid programs into national technical and financial administration structures.

Moreover, when the level of foreign financing reaches the high percentages referred to above, the consequences can be serious if donors decide to withdraw. These risks are unfortunately felt in many DCs. The 1983 report by the committee of vice-chancellors of Australian universities on the situation in the University of Pacific South (UPS), whose main campus is in Fiji, is very telling.

Besides its own budget of 10 million Fiji dollars, UPS receives an additional 5 to 10 million for foreign-financed projects each year. These sums are difficult to budget because of their unpredictability. Some of the donors by principle refuse to fund administrative costs which usually amount to at least 15%. This means that UPS has to draw this money from its own budget. Grants are also often tied to specific activities on the donors—but not on UPS—list of top priorities. The money is also very tightly controlled, in other words, cannot be allocated with any degree of flexibility. Furthermore, in many cases it has to be used to acquire inputs, be it personnel or materials, from the donor country. Donors are attracted mostly to projects that are easy to identify or see, such as a big building or a large piece of equipment that can display a plaque showing the name of the donor. Furthermore this type of a project has the potential risk of having to change size, objective, or duration. The donor even has the unilateral power to bring it to a halt, a decision that may be the result of a change of government, governmental policy or economic conditions in the donor country.

The more diversified the financing, the greater the number of potential funders, the more time has to be devoted to receiving representatives of the various donor organizations, taking them through the research centers, filling in funding applications, organizing fund management along the lines set out in the specific requirement forms and criteria papers of each of the donors, drafting midterm or final progress reports, participating in evaluation groups, etc. An African IFS grantee who became a research project administrator had the following to say: "I have had my new position for a year and a half and could now devote part of my time to research if we weren't totally submerged by the donors' requirements. At the moment I have 13 research projects going on which are financed by external aid. Keeping the donors, with their demands for reports, for meetings and their consultants satisfied, etc. takes up most of my time."

An Asian grantee who was asked about the impact of IFS financing on his research very openly said, "To the extent that I am simultaneously associated with a large number of research programs financed by diverse institutions, it is difficult for me to distinguish the contribution of the IFS and to evaluate its impact on the whole of my achievements."

A recent study conducted by the external financing service of the University of Costa Rica included a list of 72 sources of foreign financing for research for the period from 1976 to 1984 (Costa Rica 1985).

In order to study the size of the budget that the researchers in our population commanded, we asked them to evaluate their mean annual budget (not including their salary) before and after receiving an IFS research grant. Apparently, many of them found it difficult to evaluate their research budget since 118 researchers did not reply. The mean

Table 4.6. Obtaining Funds to Complement IFS Grant

Question: Has being an IFS grantee made it easier for you to obtain
other funding from

	Yes	No
Your home institution	42%	58%
Some other national institution	24%	76%
Some international institutions (other than IFS)	20%	80%

obtained for all the researchers shows that they controlled an annual
budget of U.S. $5,682 before and U.S. $13,889 after receiving an IFS
grant. Since the difference, i.e., U.S. $8,207, is slightly above the
average IFS grant, it may be assumed that the IFS grant has had a
certain catalytic effect and facilitated the obtainment of supplementary
funding. This hypothesis was confirmed by a large number of grantees, as table 4.6 indicates.

Before obtaining an IFS grant, close to 60% of the grantees controlled between nothing (19%) and U.S. $2,000; and close to 90% of
them controlled less than U.S. $10,000.

Figure 4.1 shows the distribution of the size of the survey respondents' annual budgets after receiving an IFS grant. Thus, we learned
that more than 30% of them controlled budgets of over $10,000 a year
(excluding their own salary). There are 12 researchers who administered more than U.S. $60,000 a year. Paradoxically, almost all are
Africans who are, for the most part, coordinating national programs
that are more development than research oriented. Between 40% and
90% of their total budgets are financed by their home institutions.

Let us see how this works out for our survey population as a whole.
Figure 4.2 shows that on the average more than half (53%) of the
budget available to IFS grantees comes from IFS. The second most
important contribution comes from the grantee's home institution
(27%), followed by another research-funding national institution
(10%). Only 8% of the mean budget came from international organizations (other than IFS). We know, however, that 80% of our population
did not benefit from such contributions. In fact, for half of those who
received funding from an international organization other than IFS, the
contribution they received represented between 40% and 95% of their
research budget.

This distribution of sources of funding partly explains the importance of funding from outside the home institution in the scientist's

Figure 4.1. Total Annual Budget Available to Grantee (Including IFS Grant, Excluding Salary)

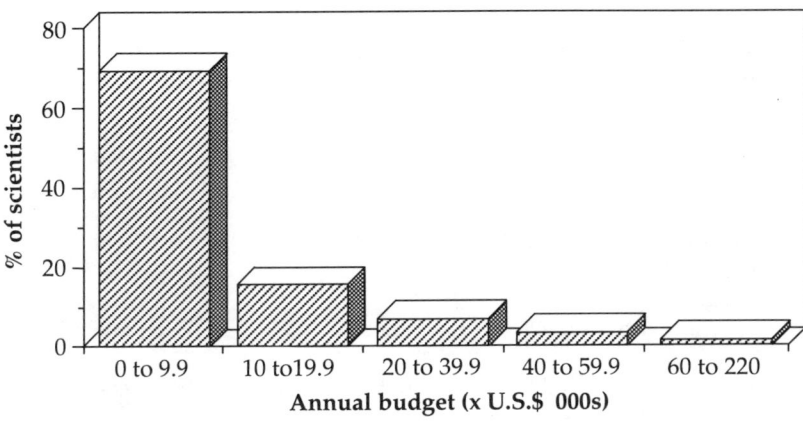

Figure 4.2. Sources of Funding

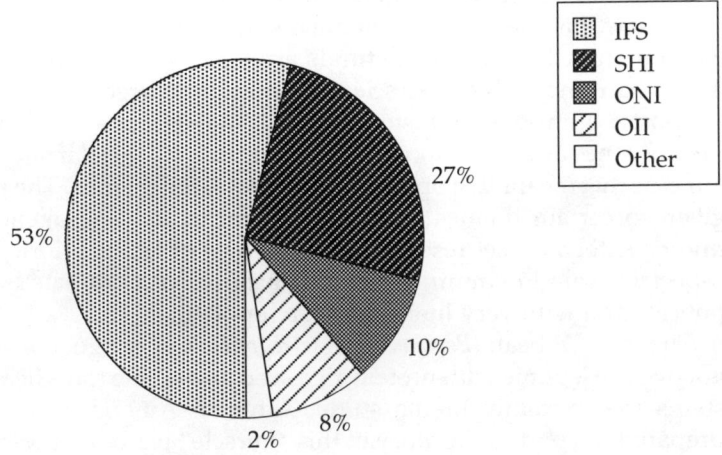

Table 4.7. Choice of Research Subject: Relative Importance of
Financing from Budget of the Scientist's Home Institution Compared
to Financing from Outside Institution

Financing by Home Institution	Essential or Very Important		Moderately to Not Very Important		Not Important At All		Total	
Essential or very important	118	48.1%	31	20.8%	11	15.7%	160	34.5%
Moderately to not very important	82	33.5%	93	62.4%	24	34.3%	199	42.9%
Not important at all	45	18.4%	25	16.8%	35	50.0%	105	22.6%
Total	245	100.0% (52.8%)	149	100.0% (32.1%)	70	100.0% (15.1%)	464	100.0%

choice of research subject. Table 4.7 shows that more than half (52.80%)
of the scientists feel that access to funding outside of their institutes is
essential or very important to their choice of research subject. On the
other hand, there were only about one-third (34.48%) that felt that
access to funding from their home institution was essential or very
important in the choice of their research subject.

We saw that the scientists' home institutions only provided slightly
over one-fourth (27%) of the funds available to the scientist and that
more than one-fifth of the scientists questioned receive no financial
support at all from their institution. This makes it possible to measure
the influence and the responsibility of outside financial institutions, in
our case that meant IFS, on the choice of research themes. The publicity
given to certain themes and their appeal to funding agencies can,
unfortunately, attract researchers to themes of lesser priority. This is
especially true in countries without nationally coordinated research
policies and with very limited research potential.

The winged bean (*Psophocarpus tetragonolobus*) is a good example. It
is a tropical legume with protein-rich seed and edible roots, leaves, and
stems that certainly merits study. After an American organization
prepared a report to the glory of this "miracle" plant, that was broadly
read in the DCs, the local reaction, in many cases, was to delve
headlong into research of unprecedented scale, without considering
the immediate ecological environment or the local eating habits. All of a
sudden everyone wanted his research program keyed toward the

Table 4.8. Possibility of Carrying Out Research Activities without IFS Funding

	No. of Researchers	Percentage
Yes, other support would have been available	72	15
Yes, even without other support	25	5
Yes, but on a reduced scale	215	45
Yes, but in a substantially different form	80	17
No	77	16
Other	12	2
Total	481	100%

generous and quick transfer of the winged bean to the rural populations. In this field, as for forestry programs—trees should grow fast, and nitrogen should rain from heaven!—science, as is often the case, is expected to produce wonders through solutions that are oblivious to the socioeconomic situation.

We see that even DC research programs are subject to fashions and that these fashions are often launched by certain research assistance organizations that have pet subjects. The DC scientists who have spent many years abroad serve as a preferred relay for the transmission of the priority themes that are proposed, *inter alia*, by these organizations, as we see from table A9 in Appendix C. Apparently, the longer a scientist has stayed abroad, the more frequently he is in contact with international research assistance organizations, besides IFS. Half of the scientists who never studied abroad were never in contact with these organizations.

In any case, the great majority of the scientists do not seem to be totally reliant on the availability of funding from IFS. Only 16% (table 4.8) of them responded negatively to the question, "Would you have pursued your research if IFS funding had not been available?" whereas IFS funding represents on the average more than half of their budgets.

Table 4.8 also shows us that the vast majority of the grantees would have been able to continue their research work in one way or another even without IFS financial support. Those who said that they would not have been able to continue their research work without IFS support were largely those for whom IFS funding represented a very high percentage of their budget. Thus, for 13% of the researchers, the IFS grant accounted for at least 90% of their research budget. At the other extreme, it is not surprising to learn that 5% of the researchers could have continued their research with no IFS support since for roughly

10% of the IFS grantees, IFS support represented less than 10% of the budget at their disposal.

Actually, the answers to this question mainly indicate that the responding scientists were prepared to change and adapt their objectives and methodologies depending on the financing but that they were determined to pursue their research, regardless of the funds available to them. The cost of research depends on methodology, equipment, and analyses. We have seen over and over again the extraordinary ingenuity of researchers in using whatever was at hand to make the instruments they needed for their analyses and measurements, as substitutes for the instruments they were not able to import. This type of an approach may not enable our responding researchers to be very competitive on the international scene; but it at least has the advantage of ensuring that they can maintain and repair their equipment, since buying sophisticated equipment is one thing, but installing and keeping it operational is quite another, as we will see in the following section.

Availability of Equipment, Vehicles, Technicians, and Scientific Documentation

It is difficult to generalize when speaking about laboratory equipment and how it functions in the DCs, for situations from one country to the next are extremely dissimilar. Within a country or even within an institution, in one place we see laboratories just as well equipped as those in the Western countries; and then, shortly thereafter, we are faced with the disheartening sight of laboratories with simple, old equipment that often doesn't work. There are very few laboratories in the DCs that have both modern, reliable equipment and the qualified permanent personnel needed to operate and maintain it. There are endless tales on how sophisticated, expensive equipment in the DCs is left idle. The electronic microscope seems to be a mythical symbol: "good research requires an electronic microscope." Unfortunately, as a result of the shortage of qualified technicians, there are very few that are in working condition.

Referring to the Universities of Ibadan and Nairobi, Eisemon told us that "at both universities, sophisticated laboratory equipment—electron microscopes, for instance—stand idle for lack of replacement, proper maintenance or the failure of the manufacturers of scientific equipment to honour their service contracts promptly" (1979, p. 518).

During my last trip to Madagascar in 1984, I saw that a nuclear magnetic resonance (NMR), that was delivered in 1981, had been out of commission for over a year. The supplier, who had been contacted immediately to repair the machine in accordance with an after-sales service warranty, had just replied that he would send a repairman in the coming weeks, without however giving any indication of the date. Since the equipment broke down, the samples to be analyzed have been sent to France, which means that each analysis requires about three months.

In his article on science in Arab countries and the Middle East, Zahlan (1970, p. 28) gives the following example. "There is much unused equipment in the United Arab Republic. For example, there are four electron microscopes which have not been used in scientific research, a magnificent mass spectrometer which lies idle, X-ray, and NMR units and numerous high resolution spectrographs which are collecting dust." It is indeed often easier to obtain costly, prestigious equipment from a donor country than glassware or reagents, or small materials that are not only less eye-catching but may even be perishable. Furthermore, the sophisticated equipment is often negotiated without the user even being consulted and without consideration for the technical implications of using it or in some cases even the import restrictions.

The example we heard about during our last trip to Costa Rica is a good illustration. L'Universidad Nacional du Costa Rica had submitted a request to the Spanish government in 1978 for various types of equipment, including an NMR for the department of chemistry. Since they received no news at all, the scientists there, after several years, lost all hope until the day early in 1985 when they received a notice from the customs office in San José stating that the NMR was theirs; all they had to do was pay the customs duties of close to U.S. $5,000. Since the university did not have that kind of money and the Spanish government did not seem to come up with a solution, the NMR, to this very day, may well still be at the airport. This type of a situation quite obviously is not compatible with the normal exercise of modern-day science.

Before even talking about delivery time for research equipment and laboratory products, i.e., often between six months and a year, we have to remember that close to one-third of the scientists in our population (and their institutions) did not have catalogs of scientific equipment suppliers; and the catalogs available to the others dated back to somewhere between 1960 and 1985. Only 15% of the scientists had recent catalogs. We also noted that when DC scientists contacted the suppliers directly, replies were few and far between. On the other hand, when IFS negotiated an order on behalf of the scientists, the suppliers telexed back a sales proposal in a day or so and, moreover, agreed to worthwhile discounts and offered attractive conditions for spare parts and delivery.

When the equipment reaches the airport or seaport, the grantee's problem is not yet over, because even if he has a letter confirming that the delivery is a gift, which should ensure exemption from customs duties, it often takes several personal visits to convince the customs officials; and, in some cases, the grantee then still has to check for any damage that may have occurred during transport. Then the equipment has to be hauled to the place where it will be used and installed. This is made easier when the instruction manual is written in a language the scientist understands and when a qualified technician is available, which is not always the case, as we will see later. One scientist we interviewed in a DC put it in these words: "In the DCs the scientists do everything and they do it alone."

The problem of under- or nonutilization of certain costly research equipment can often be traced back to the shortage of technical staff. UNESCO statistics (1985a, pp. V24–25) report the following ratio of scientists to technicians: France approximately 1 to 2, Sweden 1 to .5, Germany 1 to 1, and then, Indonesia 6 to 1, Philippines 2.5 to 1, Egypt 3 to 1. In 1975 at the Universidad Autonoma de Mexico, there were 1,885 scientists for 314 technicians, which means a ratio of 6 to 1 (Soberon and Mendoza de Flores 1985). These statistics also hide problems of job definition and evaluation. When scientists are asked to be more specific, the problem seems even worse. We asked the responding scientists whether their institutions had technicians who were capable of installing, maintaining, and repairing their research equipment. Figure 4.3 reports the results for the continents where these researchers are working.

Close to half (44%) of the institutions do not have technicians to install, maintain, and repair the research equipment. Here again Africa seems to be in the worst position since over half (51%) of the institutions on this continent do not have the technical staff required to ensure that their scientific equipment will work well; the figure is slightly over one-third (37%) in Asia.

The percentage varies considerably and depends on whether the discipline requires extensive laboratory work and research equipment such as the chemistry of crop science and microbiology. These are the fields that have the highest number of technicians available: two-thirds of the institutions working in these fields have a technician who is sufficiently skilled to repair equipment or detect a problem and identify the part that needs to be replaced. On the other hand, 61% of the

Figure 4.3. Percentage of Institutions That Do Not Have Technicians to Install, Maintain, and Repair Their Research Equipment

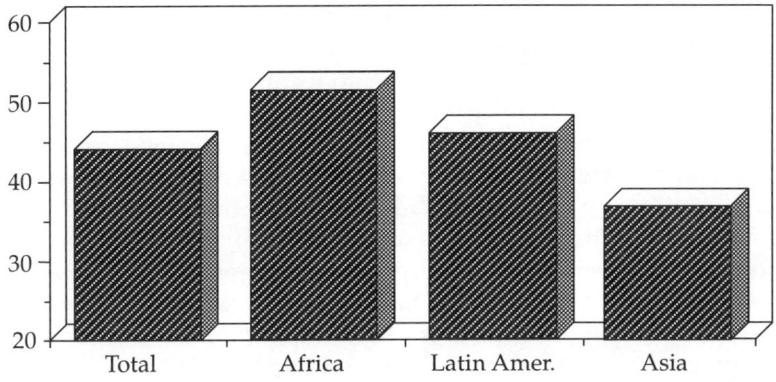

Figure 4.4. Time Required When a Foreign Technician Is Needed to Repair Research Equipment

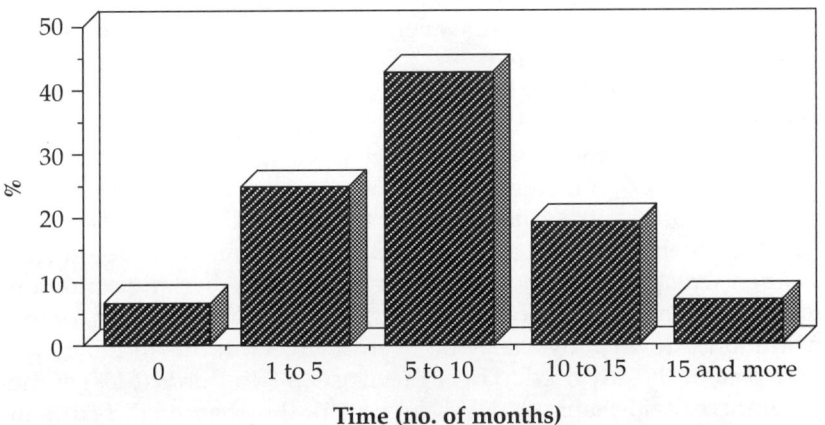

institutions working on forestry or mycorrhiza do not have any qualified technicians.

One of the consequences of this situation is that many scientists are forced to turn to foreign technicians to repair their research equipment, either by sending out the defective piece or apparatus to the supplier abroad or by having a foreign technician come to the laboratory. We asked the scientists how much time was needed for the repair. Figure 4.4 gives the answers from 214 scientists. The time varies from 0 to 84 months. Scientists who reported that their equipment has still not been repaired are not included in the figure.

More than two-thirds of the scientists (68%) had to wait 5 months or more to have their equipment repaired by a foreign technician, and more than one-fourth of them had to wait 10 months or longer. This considerable delay, of course, is a handicap for the DC scientists when compared with their colleagues in the ICs. Here again contacts made during extended stays abroad can be useful in solving the problem of broken research equipment or the shortage of equipment. Mutual assistance arrangements have been developed between some of our DC scientists and foreign institutions that hosted them during their studies abroad; these institutions cooperate, for instance, in making the structural analyses of active compounds related to natural products. This type of solution, however, does not have the makings of a satisfactory long-term system.

Another cause of delay for DC scientists is the shortage of usable vehicles. When visiting a DC research institution, it is always surprising to see that there is no shortage of vehicles, although most of them are out of order. Sometimes a relatively unimportant spare part is missing; but if it is not available locally, the vehicle cannot be used. The vehicle is then used as a source of spare parts and soon becomes a doomed wreck. Furthermore, just because it is in working condition does not mean that it will be used; there may be a fuel shortage, or the institution's travel budget may be overdrawn. With this in mind, we were pleasantly surprised to hear that close to two-thirds (64%) of the IFS grantees said that in their research work they had no difficulty in obtaining a vehicle. We should add here that IFS, by principle, does not pay for the purchase of a vehicle, so these were all vehicles bought with other funds. This is all the more surprising since in most DCs vehicle operating costs and taxes are so high that vehicles are sold for twice the price they fetch in Western countries. There was only one slight variation among the disciplines as concerned access to vehicles. Scientists working in disciplines that required the most nonurban travel, e.g., forestry, had easiest access to vehicles (72%) while scientists concentrating on laboratory research, e.g., the chemistry of natural products, had the most difficulty (60%). Many researchers, however, said they used their private car for professional travel. I was not able to measure the relative importance of this group compared to the group as a whole.

Another backbone of research is access to information. A scientist needs to stay up-to-date on scientific progress accomplished in his field. His main source of information is recent articles in scientific journals. The next chapter shows that there are other, equally valid, methods. A scientific institution, for instance, cannot be fully functional without a library or a documentation center where the scientists can do their bibliographical research. Having struggled with this problem for many years, we know that DC scientists have great difficulty in obtaining and using the scientific and technical information they need. According to B.K. Eres (1982), the weaknesses in the DC information systems can be traced directly back to national socioeconomic conditions.

The situation can be worsened by political crises of varying duration. During our last trip to Madagascar, we saw that most of the subscriptions to scientific journals stopped in 1974. That was the year that nearly all the French scientists left. We also know that exceptionally few foreign scientific journals entered Cuba in the 1960s. A journal ordered and paid for by IFS took more than one year to go from Florida to the Botanical Institute in Havana. Here again time and/or distance, combined with inadequate communications and distribution circuits, handicap the DC scientists. If they are fortunate enough to be able to subscribe to an international journal, they receive it several months after their colleagues in the ICs. Furthermore, the information contained in international journals, which by far dominate the worldwide scientific and technical information market, is not necessarily relevant nor directly usable for the DC scientists.

Only half the scientists in our reference group have access to bibliographic catalogs such as "Current Contents," and less than one-third of them (27%) have access to bibliographic data banks. And even when these two sources of bibliographic references are available the scientists cannot always locate and obtain articles or books that they have identified as perhaps being pertinent to their research.

In preparing a quantitative evaluation of the scientific information available to the IFS grantees, we asked for a list of the journals they subscribed to or could consult regularly. We have not yet analyzed the list; the numbers reported appear in table 4.9.

In more than half the cases, either the grantee or his institution only

Table 4.9. Number of Journals Researchers Subscribe to or Can Consult Regularly

No. of Journals	No. of Researchers	% of Researchers
0	45	10
1	55	12
2	52	11
3	57	12
4	52	11
5	38	8
6	26	6
7	26	6
8	16	4
9 or more	92	20

Table 4.10. Importance of Types of Publications to Research Work

Rank	Publications	Mean*
1	Foreign publications in your field	1.3
2	Books and monographs	1.9
3	Research bulletins	2.0
4	Foreign publications in related fields	2.3
5	National publications in your field	2.4
6	National publications in related fields	3.0

*Mean score based on five-point scale: 1 = essential, 2 = very important, 3 = moderately important, 4 = not very important, 5 = not important at all.

subscribes to a maximum of four journals, and 10% of the researchers do not have regular access to any journals. At the other extreme, 20% of the researchers subscribe to at least nine journals. Without a quality evaluation of the journals, we can equally assume that a good half of the researchers in our cohort are relatively unfamiliar with information produced by the international scientific community.

This does not imply that foreign journals are not important to their research, quite the contrary. No less than 96% of the researchers rated foreign publications in their field as either essential (72%) or very important (24%) to their research. National scientific journals were considered far less important (see table 4.10).

One-quarter of the scientists felt that national publications in their field were not very important (15%) or not important at all (10%) to research. Countries like China, India, Brazil, and some of the smaller DCs like Philippines and Egypt have many national scientific journals; but this is not true in many other DCs, especially in Africa. The problem of quality is more important than the problem of quantity. The conditions of acceptance for publication in national DC scientific journals do not always respect proper scientific standards. Perhaps the fact that many disciplines are not covered in DC journals further explains the greater importance the DC scientists assign to international journals. Many authors have discussed the role of national journals in DC science. Basalla (1967, p. 618), for instance, in an article on the spread of Western science in non-Western societies, went as far as to say that the creation of national journals was indispensable to the emergence of an independent scientific tradition in the DCs.

The importance a DC scientist gives to national scientific journals is probably dependent on the research and development fields they cover in his country. In a study on communications among 38 rice breeders in 10 Asian countries, Hargrove (1979–1980, p. 130) demonstrated the greater importance and relevance of the national publications. Paradoxically, in answer to a question on their choice of journals for publishing their own works, more than half answered that they preferred "international journals published in the developed countries."

It is also probable, although we have not been able to bear this out, that the longer a scientist spent in training abroad, the more importance he assigns to work published in internationally read foreign publications.

We found a close correlation between the number of subscriptions to foreign journals and the number of years a scientists spent in training abroad. This correlation is particularly visible for scientists who do not subscribe to any journals at all, as we can see from table 4.11.

Scientific journals are a necessary and important channel of communication for scientific undertakings, but not the only one. The following section explores the main modes of communication and to what extent the DC scientists use them.

Relations between Research Scientists

"Looking back on my own period of work in Lahore . . . I felt terribly isolated. If at that time someone had said to me, we shall give you the opportunity every year to travel to an active research centre in Europe

Table 4.11. Number of Scientists with No Subscriptions to Scientific Journals by Number of Years of Study Abroad

No. of Years Abroad	No. of Scientists Not Subscribing At All	Total Population	Percentage
0	11	68	16.2%
1–2	12	104	11.5%
3–4	11	113	9.7%
5–9	8	144	5.6%
10–20	1	26	3.8%
Total	43	455	9.4%

or the United States for three months of your vacation to work with your peers, would you then be happy to stay the remaining nine months at Lahore, I would have said yes. No one made the offer" (Salam 1966, p. 465). Abdus Salam shared this secret of the past with us, but it could just as well apply to many scientists in our sample today.

Many a DC scientist suffers from a feeling of isolation, especially when he has just returned from his studies in an IC and is trying to fit into the scientific community at home. Moravcsik (1976, p. 80) describes how impossible it is for DC scientists to communicate with their peers and colleagues by drawing a comparison with birds whose wings have been clipped. The feeling of isolation is probably heightened by the fact that the newly returned scientist has been trained in a large variety of universities scattered throughout the ICs. Furthermore, during this early period when the young national scientific communities are just "taking off," the scientists often have to cope with being the only specialists in their field within their institution, or even within their country.

All the authors, however, agree that science cannot exist without communications and that a colleague's criticism is vital to progress in any scientific endeavor. Here again the DC scientists are enduring a handicap little known to their colleagues in the ICs.

There are different types of scientific relations and communications. We discussed the importance of access to scientific journals. This is the most formal written form and, perhaps, the most popular form of scientific communication. Other modes of communication may be less formal but just as important and are based on interpersonal relations between scientists. They include postal contacts, telephone calls, conversations between colleagues within their institutions or during

Table 4.12. Frequency of Communication of IFS Grantees with Other Scientists (In Decreasing Order of Frequency)

Rank	Actors	Mean*
1	Scientists in your research group or department	5.4
2	Scientists in another department in your institution	3.9
3	Scientists outside your institution within your country	3.3
4	Extension staff	2.9
5	Scientists outside your country	2.8
6	A member of the IFS secretariat	2.7
7	Your thesis adviser	2.6
8	IFS grantees in your country	2.4
9	IFS scientific advisers	2.3
10	Representatives from other funding agencies	1.9
11	IFS grantees from outside your country	1.6

*Mean score based on a seven-point scale: 1 = never; 2 = rarely; 3 = annually; 4 = monthly; 5 = biweekly; 6 = weekly; 7 = daily.

travels or conferences. These forms of communication are interdependent and complementary. Most of the authors agree that discussions with colleagues are often a source of information.

To evaluate the degree of informal communication between scientists in our group, we asked them how often they interacted with a range of people, from colleagues in their home institution to foreign scientists. Answers are given in table 4.12. The figures are means for the population as a whole.

The DC scientists told us that they communicated a little less than weekly with colleagues in their research group or department. This is the same result that Busch and Lacy (1983, pp. 88–89) obtained in their study of American agricultural scientists. On the other hand, they communicated much less frequently with other scientists in their own country than did the American researchers. Communications with scientists from other institutions in their country, for instance, took place only a little more than once a year. This result may be in part explained by the fact that researchers in DCs are often the sole specialists in their field for the whole country, but this is not always the case. Our travels in the DCs often took us to institutions where we saw that scientists of the same country or even of the same institution with common lines of interest did not even know each other. In fact, there were very few who communicated more frequently with scientists from other institutions in their own country than with scientists from abroad; 35% of them only communicated once a year with scientists

Table 4.13. Frequency of Communication with Foreign Scientists as Related to Place of Education (Abroad or Home Country)

Frequency of Communication	Scientists Educated at Home		Scientists Educated Abroad		Total	
Once a month	8	12.0%	79	20.3%	87	19.1%
Once a year	26	38.8%	179	46.0%	205	44.9%
Rarely	22	32.8%	117	30.1%	139	30.5%
Never	11	16.4%	14	3.6%	25	5.5%
Total	67	100.0%	389	100.0%	456	100.0%

from their own country (other than with those in their home institution), and 42% of them communicated once a year with researchers abroad. The frequency of communication with researchers abroad is obviously related to the time spent studying abroad (table 4.13).

Thus, we were able to establish that scientists who had spent time overseas communicated more frequently with foreign researchers working in their research area than those who had not. There were almost twice as many foreign-trained researchers who communicated once a month with foreign researchers as nationally trained researchers. Correspondingly, five times more researchers who had never spent time abroad never communicated with foreign researchers.

We were also able to establish that researchers who had completed all their studies in their home country were relatively more apt to work alone on their research. A little under one-fourth of the researchers in our population worked alone, as against over one-third of the researchers who had never spent time abroad (table A10, Appendix C).

Equally, researchers who had never studied abroad were less likely to be in contact with foreign researchers, except at international scientific meetings, unlike our sample population in which close to three-fourths (74.32%) maintain such contacts (table A11, Appendix C).

During national and international conferences, scientists have an exceptionally favorable opportunity for meeting colleagues, sharing information, and discussing the progress of their work. Our own experience has confirmed that at international conferences DC scientists are underrepresented and find it difficult to command an audience when they do attend. We have to recognize that a vast majority of international conferences are held in ICs, which makes it prohibitively difficult for DC scientists to attend. The organizers of international conferences claim that it is difficult, costly, and risky to hold international conferences in the DCs. After working with DC scientists

Table 4.14. Frequency of Participation in Conferences since Obtaining IFS Grant

No. of Conferences	In Scientist's Home Country		Abroad	
	No. of Scientists		No. of Scientists	
0	97	20.6%	160	33.9%
1	68	14.4%	109	23.1%
2	73	15.5%	70	14.8%
3	61	13.0%	42	8.9%
4	44	9.3%	30	6.4%
5	32	6.8%	17	3.6%
6–10	63	13.4%	35	7.4%
Over 10	33	7.0%	9	1.9%
Total	471	100.0%	472	100.0%

in organizing conferences in many of their countries, we feel that these arguments are unfounded, especially when speaking about the Asian countries where the hospitality and efficiency are often far better than in the ICs. We have attended many poorly organized conferences in ICs we need not name.

This is a field IFS is trying to promote by backing DC scientists who organize conferences in their home country. IFS has also helped provide funds to enable grantees to participate in international conferences, but the prerequisite has always been that their work be far enough advanced to merit being presented and discussing some results. IFS has funded close to one-fourth of the participation costs for IFS grantees who have attended conferences abroad. Table 4.14 shows us how frequently the scientists in our population attended conferences since receiving an IFS grant. We see that they participated in nearly twice as many conferences in their home country (1,969) as conferences abroad (1,007). On the average the participation frequency per scientist per year was 0.84 conferences in the home country and 0.43 abroad. It should be noted that a small number of scientists participate in a large number of conferences abroad.

Thus, around 10% of the researchers participated in about half the conferences abroad, each one totaling on average 10 conferences abroad over a period of between six and ten years, in other words roughly 2 conferences per year, or four times the general average. The "champion" scored 15 conferences abroad in three years. All but one of

Table 4.15. Sabbatical Leave: Continent Where Leave Is Taken Related
to Home Continent

Continent of Sabbatical Leave	Continent of Origin							
	Africa		Latin America		Asia and Pacific		Total	
Africa	3	—	0	—	1	—	4	3%
Latin America	2	—	10	—	0	—	12	10%
Asia	0	—	2	—	7	—	9	7%
ICs	27	84%	29	71%	42	84%	98	80%
Total	32	26%	41	33%	50	41%	123	100%
Total population	82	37%	86	18%	221	45%	489	100%
% sabbatical leave/ Total population		18%		48%		22%		25%

the grantees who participated in more than 10 conferences held IFS
grants for at least nine years. They were also the most productive in the
writing field since they single-authored between 6 and 10 published
articles annually since receiving their first grant.

At the other end of the spectrum, we found researchers who were
neither very "mobile" nor very "visible," since one-third of them (34%)
had never participated in a conference outside of their country and
almost one-fourth (23%) had attended only one. Allowances must be
made for the fact that most of them had only recently received their IFS
grant. To refine the analysis would require incorporating information
on the year of the first grant.

To make science more productive, scientists and scientific data
should be constantly "on the move" from country to country. Different
systems, ranging from postdoctoral training to sabbatical leave, have
been tested in the ICs for many years already. But there are very few
DCs that, like Malaysia, have officially adopted the system of regular,
paid sabbatical leave (Thong Saw Pak 1968).

Table 4.15 indicates that one-fourth (25%) of the scientists in our
population have enjoyed at least one sabbatical leave. For over two-
thirds of them (68%), leave—which lasted between one month and
four years—was taken less than five years ago. For over three-fourths
(77%), the leave lasted no more than one year. The table shows us that
the Latin Americans benefited most from sabbatical leave (48%) and the
Africans least (18%). The proportion of sabbaticals spent in ICs (80%) is
even higher than for Ph.D. studies (75%).

At the Ph.D. level, the United States is the country that attracts the most scientists on sabbatical leave. Great Britain is second, although the number is lower than the number of students who come for doctoral studies. France, with only 9 scientists, is barely maintaining third place, which it shares with Canada. Thereafter, we have Federal Republic of Germany (6), Australia (6), Japan (4), and Spain (4).

In relating the number of years spent studying abroad to the frequency of sabbatical leave, we see that the scientists who completed all their studies in their home country are the last to take sabbatical leave; those who most frequently take sabbatical leave are 35 to 44 year old Ph.D.s who have spent between three and four years studying abroad (table A12, Appendix C).

Even if, as Charles Kidd wrote (1983, p. 409), a large proportion of the foreign scientists who are postdoctoral students or are on sabbatical leave are nothing more than "paid employees working in an American research team," they do not have the same job contract as their American colleagues and continue to be part of the staff of their home institutions that, moreover, in some cases, continue paying their salary.

The legal and employment status of scientists who accept an assignment abroad are very different. We talked to a certain number of scientists who had made this choice; all of them expressed the firm intention to return home when their contract expired. When checking, we found that indeed, except in a very few cases, they returned to their home country as intended. One-fourth (25.93%) of the scientists in our population at some time in their career had received offers to work abroad. It was interesting to note that half of the offers came from DCs, although the United States was always in first place (21%), followed by Nigeria (11%) and France (10%).

Offers were made more frequently to scientists who had spent longer periods of time abroad. A scientist, for instance, who had spent more than ten years studying outside his homeland had six times more chance to receive an offer to work abroad than his colleagues who had all their training in their home country (table A13, Appendix C). Slightly over one-fourth (27.20%) of the scientists who received offers to work abroad honored them; here again there is a clear correlation between acceptance of the job offer and the number of years spent studying abroad (table A14, Appendix C). In other words, the longer one has studied abroad, the greater the chance of receiving an offer to work abroad and the greater the tendency to accept the offer.

Chapter Five

Scientific Production

Whatever its quality, the true value of scientific research depends on its availability to the public, which is not only essential to the existence of the research scientist but also to ensuring that the results of research reach their users. There are different ways of making research public. The most common is through publication in scientific journals. Publishing is at the heart of all scientific communities and is a measure of productivity. However, it is not the only one. This chapter indicates ways to measure the production of scientists from DCs and their contribution to the reproduction of their national scientific communities. We will focus on the results of a bibliometric study analyzing the lists of publications and works by 213 scientists who answered the questionnaire sent to 489 grantees of the International Foundation for Science (IFS). As an introduction, let us look at the place of scientific production from the DCs in the world picture.

Can the DCs' Scientific Production Be Measured?

To measure the importance of scientific output from the DCs, most authors use international data bases, especially the one from the Institute for Scientific Information (ISI) in the United States (Garfield 1983, Frame et al. 1977, Blickenstaff and Moravcsik 1982). Although there is no data base that is anywhere near complete, the DCs were recently credited with approximately 5% of the world's scientific production. Many data bases are highly specialized. This is not the case for ISI, which covers some 4,500 journals from very diverse fields of science. But ISI is very selective and only screens the world's most popular scientific journals, the ones that publish the most frequently cited articles. Its Science Citation Index (SCI), developed by the ISI map makers, mainly focuses on what has become known as "mainstream science," the most internationally visible science carried in 3,100 scientific journals. Considering that there are not far from 70,000 scientific journals in the world (Turner 1984), the ISI data base is highly selective.

Thus, ISI represents about 6.5% of the scientific journals published throughout the world. Bibliometric work is often based on ISI data. Therefore, even if it covers the mainstream, it only bears on a small proportion of the world's science. Further, the DC scientific reviews are rated as "backwood cousins" in the ISI data base, which includes barely over 2% of them. French publications, together with all the other publications that are not in English, are at a disadvantage. The scanty number of DC journals, per country and per discipline, to be found in the ISI data base illustrates how severely DC science is underrepresented. The specialized review *Chemical Abstracts*, for instance, covers 201 Brazilian chemical reviews, while the SCI only covers 6 (Cagnin 1985). Out of some 200 scientific journals that exist in Thailand, ISI only covers 2 (Yuthavong 1986). For the year 1980, Argentina was only represented by 4 reviews and Mexico by 3 (Roche and Freites 1982). Fate has been no kinder to the "newly industrialized countries." ISI only covers 1 scientific publication from South Korea and 3 from Taiwan (Davis and Eisemon 1989). Out of the 76 veterinary reviews on the ISI list, only 4 come from the DCs (Russel et al. 1987). And yet numerous studies indicate that in given country-specific fields, a goodly part of the research is produced in the DCs by DC scientists, who, for instance, produce 10% of the world's scientific literature on soil sciences and agriculture (Chatelin and Arvanitis 1988a) and over 90% on cattle reproduction in the tropics (Galina and Russel 1987).

The question of adequately representing science produced in the DCs in international data bases was the main point at issue at a 1985 conference organized at ISI in Philadelphia. The title of the final conference report, "Strengthening the Coverage of Third World Science," pointed to a glaring gap (Moravcsik 1985c). It is difficult to define the precise amount of DC science omitted from the international data bases, especially at ISI. The final conference report noted that "the workshop participants estimated that only about half of the scientific output of the third world which meets international standards of excellence is included in the SCI" (Moravcsik 1985c, p. 3). ISI explains that DC scientific production published in national journals is not included in the SCI for reasons of quality. The national scientific journals are accused of not passing articles through a screening committee and publishing poor and even dubious quality work (Packer and Murdoch 1974).[1] This criticism is often addressed to India, the Third World's leading producer of science, by Indian scientists themselves (Arunachalam 1979a, 1979b, Arunachalam and Manorama 1988). The explanation often goes back to a cultural tradition that virtually bans criticism, especially in Asia. "No one wants to hurt the other. Politeness, a virtue of drawing room conversation, is extended to mean that no one criticises the other. In such an atmosphere, genuine criticism of someone's work is taken as a personal insult and leads to sentimental and emotional reactions, rather than rational defense" (Arunachalam 1979a, p. 8).

To achieve greater progress, a scientist must look critically at earlier work and, in so doing, often must criticize his elders, which is quite inconceivable in most African and Asian societies. A colleague at ORSTOM explained that he tried to work in a cooperative experiment in Indonesia but that "respect for the elders and fear of losing face made it very difficult to initiate truly scientific research. . . . since total respect for the culture was untenable, the problem amounted to finding the dividing line between what could not, and should not be changed, and what could be" (Larue 1988, p. 2).

The work published in DC scientific journals is not excluded from international science and more specifically from the SCI for reasons of quality alone. The citation criterion, which is the basis of the system, works against scientific communities at the periphery because, as we will see in greater detail below, much of the work is published in local reviews only circulated within the country. The scientists of these communities are caught in an especially vicious circle, because even when their findings are published in highly influential, prestigious scientific journals in the center, they are, all told, far less often cited than writings by their colleagues from the center (Arunachalam and Garg 1985), which explains the very ambivalent feelings of scientific communities in the periphery concerning the significance of citation. Recent work on referencing within the Brazilian scientific community showed that "citation patterns are significantly influenced by factors 'external' to the scientific realm and, thus, reflect neither simply the quality, influence, nor even the impact of the research work referred to" (Velho 1986, p. 71). Brazilian scientists feel that the place of publication strongly influences the number of times a publication is cited. This was borne out by S.M. Lawani (1977) who showed that out of a representative sample of 100 entomology articles written by Nigerian authors, articles published in foreign journals were cited 1.74 times more often than articles published in local scientific journals.[2]

Actually, as J.D. Frame so correctly said, it all depends on what you are trying to assess. "If the purpose of the bibliometric indicators is to help in the building of a national scientific inventory, telling us what kind of research is being performed at different institutions, then coverage of local as well as mainstream publications would seem important. On the other hand, if one is primarily interested in investigating Third World contributions to world science, then publication counts taken from a restrictive set would seem most appropriate" (1985, p. 121).

Thus, when Eugene Garfield (1983) prepared "Mapping Science in the Third World," he was actually measuring the impact of Third World scientific output on the international scientific community, using, as his only criterion, the part of the Third World scientific output that was cited and used by the international scientific community. For this reason it is not surprising to note that the impact was slight.

The Place of DC Science in Mainstream Science

Work by Frame et al. (1977) and more recent writings by Braun et al. (1988) reveal that mainstream scientific production is more tightly concentrated in a selected number of countries than national wealth expressed as GNP; "the gap between the haves and the have-nots is even greater in the domain of science than in the domain of economics" (Frame et al. 1977, p. 503). Ten countries produce more than 80% of the international scientific literature (84% in 1973, 82% in 1981–85). Except for India, which has maintained a steadfast eighth place since the beginning of the 1970s, all the other countries are members of the industrialized world (Frame et al. 1977). The DCs, between 1981 and 1985, produced 5.8% of the world's scientific output, of which 3.75% came from Asia, 1.15% from Latin America, 0.38% from sub-Saharan Africa, and 0.59% from the Middle East (Braun et al. 1988). Even if we challenge the representative value of these estimates, especially considering the data base used, we still see that mainstream science from the DCs is marginal compared with the rest of the world.

Within the DC group, India leads by far, producing five times more mainstream scientific publications than its runner-up of the early 1980s, the People's Republic of China.[3] Table 5.1 lists the top 15 producers of scientific literature in the DCs for 1973 and for the 1981–85 period.

This list changed considerably during the reference period. Production in certain 1973 leading countries like Brazil and Nigeria rose sharply. Some countries with small—even very small—scientific output in 1973 started climbing, e.g., Hong Kong, Saudi Arabia, South Korea. Other countries, like Iran and Lebanon, in the throes of political and military unrest, lost their standing.[4] Most of the countries on the list produced substantially more in the years following 1973[5] (Velho 1985); but the per country mainstream scientific production remained small, even in countries at the top of the list like Egypt, Mexico, and Nigeria. A comparison with the production of scientific institutions in the OECD countries shows that a country such as Egypt produces less than the Harvard University Medical School alone (Frame 1985, p. 118).

Table 5.1. Fifteen Leading DCs, Ranked by Number of Mainstream
Publications Produced

	1973		1981–85[b]	
Rank	Country	No. of Publications	Country	No. of Publications (Annual Averages)
1	India	6,880	India	10,978
2	Argentina	764	People's Rep. China	2,146
3	Egypt	683	Brazil	1,498
4	Brazil	573	Argentina	1,124
5	Mexico	368	Egypt	1,029
6	Chile	356	Nigeria	790
7	Nigeria	280	Mexico	709
8	Venezuela	200	Chile	590
9	Taiwan	186	Taiwan	509
10	Iran	174	Hong Kong	365
11	Malaysia	138	Saudi Arabia	319
12	Kenya	125	South Korea	312
13	Singapore	120	Venezuela	311
14	Thailand	117	Kenya	248
15	Lebanon	114	Singapore	214

Sources: Frame et al. 1977, table 4, pp. 507-8; [b]Braun et al. 1988.

The total production of the African continent, excluding South Africa, at present represents about one-sixth of the scientific production of a European country such as France.

Other very recent studies provide information on mainstream science production in the various Third World countries. The work done by Davis (1983) in 36 sub-Saharan African countries from 1970 to 1979 is interesting because it concerns a relatively homogeneous group of countries and shows the growing importance of universities, where 65% of the scientific output originates. Mainstream scientific production comes largely from medicine (38.2%), biology, and agronomy (approximately 22% each). Unfortunately, once again, the French-speaking countries suffer from the fact that the ISI base covers a small number of journals in French. This explains why the top six mainstream-producing countries of Africa are English speaking, viz., Nigeria, Kenya, Ghana, Uganda, Zambia, and Tanzania. Ivory Coast holds seventh place and Senegal, eighth. The leading position of the top two countries leaves no room for doubt; the rest of the list is less certain.[6]

The study of Krauskopf et al. (1986) on science in Latin America from 1978 to 1982 puts Brazil in first place, where it logically belongs, although Argentina, which is second, maintained a steady growth rate during that period. In 1978 the five leading mainstream science producers of Latin America, i.e., Brazil, Argentina, Mexico, Chile, and Venezuela, produced 92% of Latin America's total output; life sciences (medicine and biology) were the leading disciplines.

The scientific output of groups of countries (Arunachalam and Markanday 1981) and individual countries of Latin America (Krauskopf and Pessot 1983, Martinez-Palomo and Arechiga 1979, Morel and Morel 1978) has also been analyzed using ISI and other international data bases. The studies provide interesting information on the position of the various countries on the mainstream science supplier list and their impact on world science; but the description of how science is constructed in these countries, the researchers' scientific strategy, and their participation in national and international science is incomplete and often inaccurate. These studies, moreover, tend, either implicitly or explicitly, to assign research scientists of the peripheral scientific communities to two distinct categories: scientists who "really count," in other words are known to the international scientific community since they publish overseas in influential international journals, and the others, whose "local" science lacks originality and, at best, is published in low circulation local journals.

This thesis is clearly explained by Arunachalam (1988, p. 312) who spoke of scientists in India in the following terms:

Science in India appears to be divided into two distinct levels, one almost cut off from the other. At one level, where much of the good work is done, the practitioners are more at home with their counterparts elsewhere with whom they share the same invisible colleges. Naturally, they continue to publish in overseas journals, published mostly from the USA and the U.K. Rarely do they submit papers to national or local journals. Even if they submit papers to local journals, they would rarely submit what they consider to be the better ones. At the other level, many people tackle problems of not much current relevance, scientific significance or originality in the strict sense of the terms. In spite of the occasional attempts of practitioners of science at this level to publish their work in high impact journals of the West, they often have to get their papers published within India or in foreign journals of low significance. What is more, there is not so much interaction between practitioners of science at the two levels.

Several recent studies justify a revision of this exaggerated—but largely held—caricature of science production in the periphery.

Chatelin and Arvanitis (1988a, 1988b) made a bibliometric study on soil sciences and agriculture that pointed to great differences in the national and individual publication strategies in the DCs and showed that local science was not synonymous with poor science. It is not for reasons of scientific quality that the vast majority of studies on soils and agriculture are not "mainstream." Many dynamic DC scientists actually partake of the international scientific life but publish most of their findings in national journals. Studying a scientific generation's original work in this field so vital to development brought out the importance of the time needed to develop a scientific thrust.

A close look at the history of scientific production at a Mexican biomedical research institute showed that research scientists had changed their publication strategies in the score of years between 1959 and 1979 (Lomnitz et al. 1987). By 1979 half of their output was published in international journals. Yuthavong (1986), reporting on Thai scientific institutions, found a strong correlation between the number of articles the scientists from these institutions published in international journals and in the *Journal of the Science Society of Thailand*.[7]

Davis and Eisemon (1989) showed that a sizable proportion of the more dynamic scientists from four peripheral scientific communities of Asia (South Korea, Taiwan, Malaysia, and Singapore) published both in local and in international journals. The decision to publish locally is often the result of choice rather than necessity. These four countries have developed important local scientific literature that is not mainstream; and, according to these authors, the local science will probably not be eliminated as the scientific communities gain clout in the international scientific community.

All these recent findings substantiate the thesis that the bibliometric indicators, especially the SCI, do not accurately assess the scientific output from the periphery, especially from the DCs. It is impossible to measure and analyze the scientific production from the DCs since there is no operational base, no proper index of local publications and scientific work. It is, however, possible to select a series of locally published scientific reviews and compare them with mainstream science as was done by Yuthavong, by Davis and Eisemon, and by others. This is a promising but time-consuming approach since these reviews are not indexed in bibliographic bases. Furthermore, it does not account for the total scientific output, e.g., especially the unpublished work.

Our study uses a different approach. All the scientific work done by IFS grantees is covered. Our questionnaire asks about the number of articles published in scientific journals and bulletins, speeches given at

Table 5.2. Publication Production per Grantee per Year

Articles in scientific journals	1.3
Conference papers	0.5
Books or chapters in books	0.07
Bulletins and reports	0.3

conferences, books, chapters of books, abstracts, coauthored reports, etc. We also conducted a more detailed bibliometric study on publications and work by 213 grantees selected from the reference population and gave due consideration to balance between continents and between disciplines.

Total Scientific Output

Responses to the questionnaire indicate that the IFS grantees author 0.5 publications per researcher per year and coauthor 1.7, in other words, slightly more than half that of American researchers in agricultural sciences, according to Busch and Lacy (1983), who reported 0.9 and 1.3, respectively.

Using lists of publications and working papers gives a slightly higher figure, i.e., 1.3 journal articles (sole author and coauthor) instead of 1.2 (0.5 + 0.7), as in the preceding case (table 5.2).[8] Further, we have been able to estimate that more than half (55%) of the total scientific production of the grantees was published or available locally. The remaining 45% that was published abroad can be divided into articles published in scientific journals in ICs (37%) and in other DCs (8%). These global statistics camouflage significant variations between geographical and scientific areas, as table 5.3 shows.

The field in which the IFS grantees are publishing most (1.6 publications per grantee per year) is natural products. This is also the field in which they publish most abroad (1.1 publications per grantee per year). Food sciences is almost the opposite. There are more local (1.0) publications than foreign (0.4) publications. These results can be traced to the nature of the related research. The fields in which there are the fewest publications, i.e., forestry (0.7) and rural technology (0.8), are probably also the fields with the most practical applications, whose results do not always need to be published.

We have also observed that Asian grantees publish more than African or Latin American grantees (1.5 as against 1 journal article per grantee per year, respectively). Further, Asian grantees publish more

Table 5.3. Number of Journal Articles Grantee per Year by Discipline

Discipline	Published Locally	Published Abroad	Total
Aquaculture	0.6	0.7	1.3
Animal production	0.8	0.4	1.2
Crop science	0.5	0.6	1.1
Forestry	0.4	0.3	0.7
Food sciences	1.0	0.4	1.4
Natural products	0.5	1.1	1.6
Rural technology	0.4	0.4	0.8
Total mean	0.64	0.66	1.3

Table 5.4. Place of Publication per Geographical Area (%)

Geographical Area	Locally	In Another DC	In an IC
Africa	41	10	49
Latin America	58	9	33
Asia	60	6	34
Average	55	8	37

locally (60%) than African grantees (41%). In Latin America more was published locally (58%) than overseas (table 5.4). These percentages, in comparison with figures for developed countries, are exceptionally high. Scientists in France publish 20% of their scientific production in foreign journals. For Western Europe as a whole, the figure is 12% and for Japan 25% (Garfield 1977, 1978, 1983).

These figures cover the grantees' total scientific production, not only journal articles that are published in equal proportions in local and foreign journals. When consulting table 5.4 remember that there are many more local journals in Asia and in Latin America than in Africa.

We have also observed a relatively significant difference in productivity by gender; men publish more than women (table 5.5). This difference is all the more pronounced since women are more active in much published fields such as food science and natural products and less active in fields such as rural technology where little is published. Women tend to publish more locally than men.

Table 5.5. Number of Journal Articles per Grantee per Year by Sex

Gender	Published Locally	Published Abroad	Total
Men	0.65	0.73	1.38
Women	0.53	0.33	0.86

Table 5.6. Number of Publications (Including Bulletins, Books, Internal Reports, Conference Papers)

Research Area	As Sole Author	As Co-author	Total
Aquaculture	0.9	1.3	2.2
Animal production	0.4	1.6	2.1
Forestry	0.7	1.2	1.9
Food science	0.9	1.7	2.6
Natural products	0.4	1.9	2.3
Rural technology	0.7	0.9	1.6
Average	0.7	1.4	2.1

Research is becoming increasingly collective, and scientists work together not only to bring their research to a successful conclusion but also to be able to publish their results under joint authorship. IFS grantees, for instance, publish about two-thirds of their work with coauthors, as is shown in table 5.6.

Table 5.6 illustrates the general rule that the fields in which scientists work together most are the fields in which most is published. This confirms earlier findings by de Solla Price and Beaver (1966) and by Beaver and Rosen (1978, 1979a, 1979b) who observed that collaborative research enhanced productivity.[9] Results also show that there is significant difference between disciplines. Fields that have the largest number of authors per publication, such as natural products, are fields that require inputs from a variety of disciplines, e.g., taxonomy, botany, chemistry, and pharmacology. If the right specialists are not locally available, foreign cooperation is required, which explains the higher number of foreign coauthors per publication (0.53) for a field such as natural products (table 5.7). Although the difference in average numbers of coauthors in terms of geographical distribution is not signifi-

Table 5.7. Average Number of Authors and Coauthors (Local and Foreign) per Publication

Research Area	No. of Authors	No. of Local Authors	No. of Foreign Coauthors	Total No. of Publications per Grantee per Year
Aquaculture	1.87	0.75	0.12	2.2
Animal production	2.12	0.98	0.14	2.6
Crop science	1.95	0.72	0.23	2.1
Forestry	1.98	0.67	0.31	1.9
Food science	2.12	0.98	0.14	2.6
Natural products	2.85	1.32	0.53	2.3
Rural technology	2.20	0.80	0.40	1.6
Average	2.25	0.96	0.29	2.1

cant, we have noted that Asia has the highest number (2.4), followed by Latin America (2.2) and then Africa (2.1).

The mean number of authors per publication gives an interesting indication of the degree of association among researchers who publish, and the origin (local or foreign coauthors) gives an indication of the openness and/or dependence of the researchers. Table 5.7, for instance, confirms that natural products is the field for which the publication rate is the highest. It is also the field that brings IFS and foreign scientists together most. Actually, the more the scientists publish abroad, the more they work with foreign scientists in their preparation work.

Thus, we found that there were no researchers who had published more than 12 articles abroad without a foreign coauthor. Garfield (1983) has shown that articles by researchers in DCs have a greater impact (on the international scientific community, measured in terms of number of citations per article) when they are coauthored by researchers from ICs. Here we come up against the dilemma of the strategic scientific choices that researchers in DCs, in common with most researchers in peripheral scientific communities, have to make between participation in mainstream science (the most used, most visible, and most frequently cited) and the resolution of local problems through "inward-looking" research, which some call "domestic" or "inbred" science.

It is worth observing that coauthoring with foreign scientists is the most prevalent among scientists who studied or went on postdoctoral study tours abroad. In most cases, however, these publications are

Table 5.8. Language of Publication by Linguistic Area

Linguistic Area	Local	English	French	Spanish & Portuguese
French-speaking countries	1	17	82	—
English-speaking countries	8	92	—	—
Spanish- & Port.- speaking countries	—	36	1	63
Average	6	76	8	10

produced in the years immediately following the stay abroad; sustained active collaboration is rare. Other associations develop when a foreign professor is on assignment in the scientist's home laboratory or when expertise, not locally available, is brought in as part of a program financed by a foreign institution or as follow-up to an international conference.

The choice of language of publication is also central to the scientific strategy. A look at the lists of references consulted in preparing this study confirms the hypothesis that the different linguistic worlds are almost "language proof," especially between the English and French languages. Spanish and Portuguese speakers often cite literature in English; this is rarely the case for French speakers. And references by English-language scientists are drawn, for all intents and purposes, exclusively from literature written in English (table 5.8). To one degree or another, these four languages dominate the world's scientific literature. In a few Asian countries, science is published in national local languages.

The percentages in table 5.8 refer to 5,000 publications produced by 40 Latin American researchers (mainly Spanish speaking), 29 French-speaking African researchers, and 138 English-speaking researchers. These results confirm the prime importance of English and the resulting subordination of the other languages. Out of 3,678 publications by English-speaking grantees, 2 were in French, 1 in German, 4 in Russian, and none in either Spanish or Portuguese. On the other hand, more than one-third (36%) of the publications by Latin American and almost one-fifth (17%) by French-speaking grantees were in English. Our case study in Senegal showed that English was increasingly used in French-speaking countries. The percentage of articles published in English by scientists working in Senegal, for instance, rose from 15% in 1975 to some 30% in the early 1980s.[10]

The other conclusions that can be drawn concern the relatively significant use of local languages in certain Asian countries, e.g., Indonesia where more than half (52%) of the grantees' published works appear in Indonesian languages, Thailand (28% in Thai), and South Korea (18% in Korean). These percentages would be considerably higher if our figures only applied to the language of publication used in the national journals. Eisemon and Davis (1989) reported that over half (57.1%) of the articles in six South Korean journals were published in Korean.[11] Publication strategies differ greatly, depending on both the country and the discipline. Unlike South Korea, in Singapore all the scientific journals are in English.

As a general rule, researchers will tend to write in local languages for national publications if their subject of research is for direct application. Except, perhaps, for a few Thai scientists, who find it difficult to write in English, the decision to publish in a national language and in a national publication is usually a question of strategy, as can be seen in interviews with scientists who say things like, "I submitted this paper to a local journal because the contents essentially bear on a local problem. This should make it easier for me to make the authorities aware of the problem and help them find the right solutions for our national development." Or, as concerned Korean and Thai, "I published in my national language so that I could use it in teaching."

Another scientist said that he had decided to publish in a new local journal to contribute to its development because "I feel that it is essential for our countries to have good quality scientific publications."

A few scientists admitted that it was "easier" and "faster" to publish in national journals. Using a national language also means reaching readers who do not receive international journals and, furthermore, gaining repute among peers and students in the home institutions. Most published in both national and international journals. Only about 20% published exclusively in the national journals; these were mainly young scientists working in agronomic research (animal production, crop science, forestry) and aquaculture. There were no scientists from the natural products group who published exclusively in the national journals.

An analysis of the references used in articles provides precious information on the scientific output and research practices. We saw, for instance, that there was far more intralinguistic than interlinguistic transfer. We also obtained information on the relative use of local and international science and the relative age of the work cited in the scientists' publications. Some authors found that scientists from the peripheral countries tended to ignore—or did not have access to—older publications and thus concluded that the use of more recent

Table 5.9. Breakdown of Age of Reference Cited by Continent of Scientists' Work

Years	Africa		Asia		Lat. Am.		Total DC		Center Countries	
0–5	180	22%	195	22%	144	25%	519	23%	340	42%
6–10	312	38%	240	27%	180	32%	732	32%	239	29%
over 10	327	40%	456	51%	240	13%	1,023	45%	232	29%
Total	819	100%	891	100%	564	100%	2,274	100%	811	100%

Table 5.10. Breakdown of Age of Reference (in Years) Cited by Scientific Discipline

Discipline	0–5		6–10		Over 10		Total
Aquaculture	132	23%	138	24%	303	52%	573
Animal production	42	17%	96	38%	114	45%	252
Crop science	102	18%	171	31%	285	51%	558
Forestry	51	26%	75	37%	75	37%	201
Food sciences	48	21%	108	47%	75	32%	231
Natural products	144	31%	144	31%	171	38%	459
Total	519	23%	732	32%	1,023	45%	2,274

references was characteristic of science in these countries (Rabkin and Inhaber 1979). The opposite was also alleged, i.e., that scientists in the peripheral countries cite references that are much older than those cited in articles published in international journals by colleagues from ICs (Arunachalam and Markanday 1981; Velho 1985, pp. 244–56).

Now let us look at our study population. For purposes of comparison with scientists from developed countries working in similar fields, we referred to Lea Velho's thesis (1985, p. 247) to find a sample of articles published by mainly American scientists in scientific journals of center countries. The results (tables 5.9, 5.10, 5.11, and 5.12) show that DC scientists generally refer to articles more than 10 years old. Close to half (45%) of the references date back to over 10 years, while for authors from center countries the figure is under one-third (29%).

Scientists from center countries often (42%) use references under 5 years old, while for DC scientists the figure drops to 23%. Table 5.9

Table 5.11. Breakdown of Age of Reference between Publications
Published Abroad and Nationally

Years	Abroad		National		Total	
0–5	405	25%	114	17%	519	23%
6–10	537	34%	195	29%	732	32%
Over 10	660	41%	363	54%	1,023	45%
Total	1,602	100%	672	100%	2,274	100%

Table 5.12. Breakdown of Age of Reference, Foreign vs. National

Years	Foreign Ref.		National Ref.		Total	
0–5	243	14%	276	56%	519	23%
6–10	606	34%	126	26%	732	32%
Over 10	936	52%	87	18%	1,023	45%
Total	1,785	100%	489	100%	2,274	100%

shows that there was no great difference between geographical areas
for the three main continents although there were differences between
disciplines, as has been shown in other studies (Crane 1972)[12] and in
table 5.10.

The figures indicate that natural products, a discipline that draws
heavily on organic chemistry and pharmacology, uses the most recent
references (31% within the last 5 years). It is worth remembering that
this is the field that generates the most joint publications with foreign
researchers. The biological sciences most directly linked to agriculture
(animal production and crop science) and aquaculture are the disciplines with the most references over 10 years old (between 45% and
52%). The findings concur with Weiss (1960) who showed that biological disciplines, largely based on analytical work, e.g., natural substances and work on mycorrhiza in forestry, tend to use more recent
references than the more descriptive research that relies more on
experiments with live matter.

As concerns the age controversy with regard to "national versus
international" journals, our results (table 5.11) tend to agree with
Arunachalam and Markanday (1981). Apparently articles published in
national journals cite references that are older than those cited in
international journals that belong to mainstream science. A finer analy-

sis would probably reveal significant differences between countries. Eisemon and Davis (1989) showed that one-fifth of the references in articles published in national journals of Malaysia, Thailand, and South Korea dated back to 5 years ago at most while in Singapore nearly one-third of the nonmainstream science references were of that age.

It is quite clear that articles printed in national reviews are much more readily assimilated by DC scientists than anything found in foreign journals, as table 5.12 indicates.

Over half (56%) of the references drawn from national scientific literature date back at most 5 years, while only about one out of seven references (24%) taken from foreign journals is thus dated. Yet the scientific transfer within or between the DCs is not very great (only 22%). In other words, references in publications by DC scientists are mainly (78%) taken from mainstream scientific literature but with some delay since more than half the references date back to at least a decade ago.

Several reasons can be suggested for this situation, which is largely due to dysfunctioning of scientific practices in the DCs. Since most of the DC scientists, unlike their colleagues in developed countries of the center, do not belong to what is generally called the "invisible college," they do not become familiar with their colleagues' work before it is published. Actually, their only access to information is tedious bibliographic research, and even this does not always result in the identification of the most relevant reference work. In the preceding chapter, we saw that only half the scientists had bibliographic catalogs like "Current Contents" and that less than one-third of them had access to bibliographic data banks. The unavailability of bibliographic references was felt with special acuteness in Africa.

This said, during our visits we saw that, except in a certain number of African countries, the libraries in DC universities and institutions had relatively recent scientific journals from countries of the center that institute scientists rarely consulted. Some of these journals looked as if they had never been opened. Many scientists try to subscribe individually to the most relevant international journals, but scanty, irregular financial means makes this difficult.

The fact that DC scientists often cite articles in journals that are over 10 years old can also be related to the time between their training period abroad and the publication of their work. Over 75% of our cohort studied abroad, mainly in the United States, Great Britain, and France. Quite possibly their references are works they learned about during their education abroad. This is an explanation Lea Velho enter-

Table 5.13. Receipt of Prizes or Awards by Number of Papers
Published

No. of Publications	0	1–9	10–20	21 and Over	Total
Grantees who obtained awards	20	28	17	9	74
%	8	15	49	60	16
Total no. of grantees	238	183	35	15	471

tains in reference to Brazilian scientists: "The longer the time since the
researchers returned to Brazil from graduate training abroad, the older
the foreign literature they tend to cite" (1985, p. 253).

Turning to total scientific production, we see that English-speaking
grantees, mainly in Asia, constitute the most published group (2.37
publications per scientist per year), while French-speaking Africans
(1.63) and Latin Americans (1.76) form the least published groups.
These figures, of course, only provide a part of the picture and cannot
be used as a definitive measurement of the quality of a research
scientist. Other indicators have to be used. For reasons given above, we
decided not to use the citations method in measuring the impact and
the quality of articles published in international reviews. A full-fledged
qualitative evaluation would have required the participation of several
specialists with a variety of linguistic capacities for each of the disciplines concerned, which was beyond the means of our study. To
conclude this chapter on scientific production, we looked at the scientific tribute bestowed on the grantees in recognition of their contribution to their respective fields.

Scientific Recognition

For the period between 1974 and 1984, in recognition of the quality of
their scientific work since obtaining an IFS grant, 74 grantees, i.e.,
about 1 out of 6, received at least one scientific award. The frequency of
tribute is clearly related to the year of first grant, since rewards (one or
more) were meted out to some 1 out of 4 grantees who obtained their
first grant in the 1974–79 period and 1 out of 3 for the 1974–75 period.

We also found links between scientific awards in DCs and the
number of publications credited to a scientist, although 20 of the 74
recipients had not had anything published, as table 5.13 shows. Thus,
we can see that more than half (60%) of the grantees who published at

least 21 times received some credit. Over two-thirds of all forms of tribute were extended to Asian scientists; the rest were divided equally between African and Latin American scientists.

Most prizes were national expressions of merit, such as the Science Academy Medal for Young Scientists (India), the Outstanding Young Scientist's Award (Philippines), the Gold Medal of the Pakistan Academy of Science, Best Woman Scientist of the Year (Sri Lanka), or the First National Prize for S&T Research (Congo), or awards from national scientific institutions, such as the Hooker Award (Indian Agricultural Research Institute, New Delhi) or the Best Scientist of the Year (Mahidol University, Thailand). Far less often, the prizes or recognition were international, such as the United Nations University Prize (biological sciences), the IBPGR/FAO Prize (genetic resources), Fellow, Third World Academy of Science, or the Guinness Prize of Scientific Merit for Development.

Production of University Graduates

We know that more than two-thirds of the grantees (71%) work in universities or higher education establishments and devote much of their time to teaching. The teaching function is essential not only to train senior civil servants and to reproduce the scientific community but also for research purposes proper. All too many university professors in DCs have lost contact with ongoing science and merely chant scientific facts of the past instead of teaching students that science is an active method used to state and solve problems. The very concept of higher education demands that teachers participate in research and similarly that research scientists at some time in their career contribute to the teaching world.

This is the case for the majority of IFS grantees. We asked them how many students they had trained since receiving an IFS grant. The answers appear in table 5.14.

The number of people trained is, quite understandably, inversely proportionate to the level of the training. It is interesting to note that during the first 10 years of IFS there were practically as many Ph.D.s trained by IFS grantees as there were grantees, although only one-quarter (26%) were actively involved. More than half (54%) of the grantees were involved in M.S. training. Many of the students, of course, benefited from the financial support and equipment obtained by the IFS grantees, as we can see from table 5.15.

Thus, a considerable majority (62%) of IFS grantees make the equipment purchased with the IFS grant available for use in teaching ac-

Table 5.14. Number of People Trained since Becoming IFS Grantee

	No. of People Trained	No. of grantees Involved in Training		No. of People Trained per Grantee
Postdoctorate	138	30	(6%)	0.06
Ph.D.	484	130	(26%)	0.98
M.S.	3,242	262	(54%)	6.60
Technicians	3,194	305	(62%)	6.50

Table 5.15. Percent of IFS Grantees Who Made Equipment Bought with Grant Available to Others

	Yes	No
Teaching activities?	62%	38%
Other research projects in their research unit or department?	90%	10%
Other research projects in other departments of their institution?	56%	44%
Other research projects in other institutions?	25%	75%

tivities, and far more (90%) make it available to other projects in their research unit or department. These percentages diminish as a function of geographical distance, but it is interesting to see that one-fourth (25%) of the grantees share the equipment acquired with IFS support with research projects in other institutions. This must be taken into account in any attempt to evaluate the impact of a research assistance institution such as IFS in DCs. IFS, through its grantees, unquestionably helps a whole network of DC scientists, well beyond the grantee's nearest colleagues; but it is difficult to produce numbers.

Implementation of Research Results

Research is not an end in and of itself. A nation, like a business, that only works to produce research results would actually be working for its competitors. The video tape recorder, for instance, was invented in the United States but was manufactured and sold by Japan. Two recent

Table 5.16. Implementation of Research Results by Geographical Area

Geographical Area	No. of Projects Implemented	Total No. of Scientists	%
Asia	76	221	34
Africa	49	182	27
Latin America	22	86	25
Total	147	489	30

studies by the U.S. Congress Joint Economic Committee show that the drop in American productivity in the 1970s could be traced back to misguided and, even worse, badly used R&D. In the DCs, proper use of the research results is of great concern to political authorities and funding agencies, since research all too often has little effect on development except in fields like agriculture and health. In the ICs, most of the resources devoted to R&D are intended to boost production. In the DCs, the situation is the opposite. Because of a stringent social, economic, and political context, it is very difficult to ensure broad dissemination of research results and technological progress.

We do not intend to expand on the main counterpoint theories involved in making and spreading technological progress. We merely want to look at the scientists' perception of the problem of applying the research results and, through their replies to the questionnaire, assess relative success. But until these early results are further checked and researched, they must be interpreted with caution.

Close to one-third (30%) of the grantees said the results of their research had been, or were being, applied. Percentages recorded differed according to continent, the number of grants obtained, and research discipline (table 5.16). It is in Asia that we found the most project results being applied (76 out of 221). Thus, slightly over one-third (34%) of the projects carried out by Asian grantees led to applications, as against roughly one-fourth for African (27%) and Latin American (25%) grantees.

The greater the number of grants, the greater the number of projects whose results led to some degree of application. Further, judging from the results set out in table 5.17, certain fields of research progressed into the practical application phase more easily than others. There were proportionately about three times more projects that led to applications in the field of aquaculture (41%) than in the least applied field of natural products (13%). Yet this latter field generates the highest number of publications, including publications in foreign journals. The fact that

Table 5.17. Number of Projects Implemented Per Research Area

Research Area	No. of Projects Implemented	Total No. of Grantees	%
Aquaculture	40	98	41
Animal production	22	75	29
Crop science	41	113	36
Forestry	9	39	23
Food science	18	53	34
Natural products	12	90	13
Rural technology	5	21	24

Table 5.18. Beneficiaries of Implemented IFS-Supported Research

Rank Score	Mean Score*		Rank Score	Mean Score
1	3.52	Other scientific disciplines	1	4.26
2	3.29	Small farmers	2	3.92
3	3.25	Rural residents	3	3.82
4	2.93	Local or state govt. agencies	7	3.34
5	2.73	Large farmers	4	3.56
6	2.64	Agribusiness	5	3.46
7	2.63	General public	9	3.20
8	2.49	Urban residents	6	3.41
9	2.42	Local industries	8	3.25
10	1.87	Foreign or int'l industries	10	2.61

*Mean score based on five-point scale: 1 = not at all; 5 = a great deal.

aquaculture easily leads on the honor roll confirms certain results reported in a survey carried out in 1980 by Berlinguet (1980).

Out of 10 projects that seemed the most likely candidates for practical application, 5 were in aquaculture. Promise in this field seems to be derived from the fact that experimental conditions are often similar to real production conditions or can be transferred at an investment cost that soon translates into higher profits. Further, the grantees often serve as advisers to the producers.

We also wanted to know who benefited from the research results obtained by IFS grantees. To do so, a series of past, present, and

Table 5.19. Implementation of Research Results: Vectors and Actors

	No. of Projects Implemented	%
Public sector (mainly extension services)	73	50.0
Farmers	23	15.9
University research institutes	15	10.0
Private sector	15	10.0
Cooperatives	8	5.4
Publications	7	4.7
International Organizations	6	4.0
Other	1	—

potential beneficiaries was identified and ranked on the basis of answers provided by the grantees (table 5.18). It would be premature to analyze the table; but it is interesting to note that the grantees felt that their colleagues from other disciplines were, and will continue to be, the primary users of the results of their research. Neither national or international industry seems to pay much heed to this research. In between, the rural populations, especially the small farmers, seem to benefit most and will continue to do so.

Again, according to the grantees, application of the results of research depends, and will continue to depend, on the public sector, mainly the extension services. Table 5.19 further evidences this. The grantees themselves felt that international organizations played a very minor role in this domain. We do not have complete information on interaction with other international organizations; but we have ascertained that in most cases their support was designed to provide IFS grantees with the financial, material, and manpower backing needed to integrate the IFS project into a much broader multidisciplinary research program by including a training component and external partners or to start up a new research program not connected to the IFS project. This especially applies to organizations such as U.S.AID, BOSTID (Board on Science and Technology for International Development), and IDRC. It may be profitable to think about creating an international body to work on the application of research results produced by scientists in DCs.

Several conclusions can be drawn from this chapter concerning the specific nature of science produced by DC researchers and the construction of science in their countries.

Quality is not the only reason why science produced in DCs is not adequately represented in international data bases. Some, like the one at SCI, are not properly equipped to evaluate the scientific production from science communities in peripheral regions, in particular in the DCs. They can be used as a source of information of the relative strengths of various countries in mainstream science and their impact on world science but give an incomplete and often inaccurate picture of total scientific output and how science is constructed in nonmainstream countries.

A look at total scientific production shows that DC scientists often publish (up to 60% in Asia) in national journals, that the leading language is English, a language even used for publishing by close to one-fifth of the French-speaking scientists and over one-third of the Latin American scientists. We also saw that the English-speaking scientists only publish in English or, as is the case in Asia, (often) in local languages. Most of the scientists publish in both national and international journals. Although publication strategies differ according to country and to scientific discipline, scientists who decide to publish in a local language or journal do so by choice and not by necessity.

DC scientists cite references essentially (78%) from mainstream scientific literature that they seem to receive later than their colleagues in the center since nearly half the references are over 10 years old, as against 29% of the references cited by scientists from the center countries. An analysis of the citations indicates that DC scientists use more recent articles from national journals than articles from international journals. Actually, citation modes are affected significantly by factors unrelated to science, factors that are social rather than cognitive in nature. Scientists in the DCs have little or no informal contact with their colleagues in developed countries, and they need much more time to avail themselves of new scientific data that are pertinent to their research.

In sum, DC scientists often cite their colleagues from the developed countries, but their own work being relatively "invisible" is seldom cited. They often feel caught in a dilemma: either adopt the habit of scientists from developed countries and publish in international journals to become more "visible" and gain international standing or else seek national recognition by publishing in local journals and sometimes in local languages, thus being condemned to nonexistence or, at best, marginal existence in mainstream science. The general trend is to adopt the two strategies together.

Considering the dilemma, and since the circulation of scientific information is notoriously out of kilter, it is impossible to judge the quality of scientific work exclusively on the basis of citations. For this reason, we must reject the abusive simplification that predicates on the citation count criterion and seems to consider local science as a synonym of poor science and international science as a synonym for good science.

1. We know that most of the scientific journals published in the DCs are short of articles and many seldom reject any, but we also know that the editorial practice of certain mainstream journals, including some of the leading ones, is not always very selective. Packer and Murdoch (1974) asserted that during the 1963-73 period, the *Bulletin of the Entomological Society of America*, by principle, and insofar as possible, printed all the articles it received. During that decade only 4% of the articles submitted to the *Journal of Economic Entomology*, the *Annals of the Entomological Society of America*, the *Bulletin of the Entomological Society of America*, and *Environmental Entomology* were rejected.

2. In this publication Lawani also provides a per country rundown of the 829 journals that have an above-average impact according to the SCI. The United States is the leader (60% of the titles), followed by Great Britain and the Netherlands. There is only one DC journal (*Revista Mexicana de Astronomia*, published in Mexico) on the list.

3. The scientific output from the People's Republic of China was completely ignored in the 1973 SCI base (except for one publication). The powerful emergence of this country at the end of the 1970s was due to three interdependent phenomena, viz., increased contact with Western science, a sharp rise in the number of scientific journals published in the country, and ISI's stated decision to correct the under- or nonrepresentation of that country in the ISI data base.

4. Thailand and Malaysia, which had disappeared from the list of the top 15, were sixteenth and seventeenth in the 1981-85 period, with annual average production figures of 188 and 169 publications, respectively.

5. Brazil's progression has been especially remarkable considering the highly internal focus of its scientific output. In the field of agricultural research, Brazilian scientists publish more than 90% of their results in national journals in Portuguese. See the excellent Ph.D. thesis by Lea Velho entitled "Science on the Periphery: A Study of the Agricultural Scientific Community in Brazilian Universities" (University of Sussex 1985).

6. The PASCAL base (applied program for automatic compilation and selection of literature) created by CNRS in France contains approximately four times more references on Senegal that ISI does.

7. The *Journal of the Science Society of Thailand* is also indexed in the SCI of the ISI. A weaker correlation was noted between the number of publications appearing in the latter's journal and the number of abstracts presented at the

annual symposium of the Science Society of Thailand by the various scientific institutes of Thailand.

8. In a more precise quantitative and qualitative evaluation, allowances have to be made for the relative value of the various types of publications. What value should be assigned to a book or a chapter in a book as compared to an article published in a scientific journal? Should the value of a scientific article be considered inversely proportionate to the number of its coauthors?

9. Three reports by Beaver and Rosen published as a series in *Scientometrics* in 1978 and 1979 are based on a study of collaboration between scientists through time since the seventeenth century. This study showed that collaboration in scientific research was a sign of professionalism within the scientific community and made the scientists more mobile and "visible."

10. The case study on the history and the development of the scientific community in Senegal, presented in the second part of my thesis, is also relevant here (Gaillard 1989).

11. The fields covered by these journals, viz., biology, biochemistry, computer sciences, electronics, and physics, are directly related to international science.

12. Brown, cited in Crane (1972), found that the percentage of references under 10 years old was the highest in publications on physics, lowest in biology and that physiology and chemistry were in between.

Chapter Six

National Scientific Communities: Costa Rica, Senegal, Thailand

In the preceding five chapters, we repeatedly pointed out that global statistics blurred regional disparities and dissimulated very pronounced differences between DCs, even those within the same region. In any case, there is no single Third World; there are Third Worlds, with "levels of under-development no less unequal than levels of development" (Salomon 1984, p.46). We felt that a comparative study of research scientists and the emergence of national scientific communities in several countries might better explain disparities and correlations of situations and problems and also consolidate results from our survey of the scientists. Chapter 6 abridges results from this comparative study; the complete results are reported in my Ph.D. thesis (Gaillard 1989).

Despite their incongruity, the countries called the Third World can be seen as a set, broken down into subsets with some degree of uniformity in size, industrialization, position in the world market, demography, wealth, dependence on natural resources, and the often precarious balance between natural resources and population pressure. The most common typologies are linked to economic indicators, especially the per capita gross national product (GNP) and suggest a classification based on thresholds, e.g., the World Bank typology that recognizes low-income countries (U.S. $0–400 per inhabitant per year), medium-income countries (U.S. $400–1,700), and oil exporting high-income countries (more than $1,700). In the United Nations systems, especially UNCTAD (United Nations Conference on Trade and Development) makes a distinction between several categories: newly industrialized countries (NICs), oil exporting countries, and the least developed countries (LDC). But the search for a more "explicative" typology must extend beyond quantifiable factors to include social structures, political systems, and national history.[1]

Several criteria guided our choice of countries for this study. First, we wanted to choose at least one country from each of the three major continents of the Third World. For practical reasons (lack of time and resources), we decided to take small- or medium-sized countries that

Table 6.1. Brief Comparison of Country Characteristics

Country	No. of inhab. 000s[a]	Area km^2	Density inhab/km^2	GNP U.S.$/ inhab[b]	Urbaniz. rate
Costa Rica	2,613	51,100	51.1	1,340	50.3%
Senegal	6,793	196,722	34.5	370	25.4%
Thailand	53,722	513,115	104.7	810	19.8%
France	55,623	543,964	102.3	9,545	73.4%
United States	243,773	9,529,063	26.0	17,600	73.7%

Source: "Les Chiffres du Monde," Encyclopedia Universalis, 1988.
[a]1987
[b]1988

showed signs of emerging national scientific communities. This explains why we automatically excluded countries as big as continents such as China, India, and Brazil, which, as we saw in the preceding chapter, are the three major scientific powers of the Third World. In order to avoid having to compare countries with levels of development too far apart, it seemed wise to avoid NICs such as the four Asian "dragons" (Taiwan, South Korea, Singapore, and Hong Kong) and certain Latin American countries (Argentina and Mexico). Our choice was also guided by ease of access to information and effective cooperation with national scientific leaders. The result was that we chose Costa Rica to represent Latin America, Senegal for Africa, and Thailand for Asia. Table 6.1 briefly describes these countries in relation to France and the United States by indicating size (area and population), wealth (GNP/per inhabitant), and urbanization rate.

The three countries in the study are small (Costa Rica and Senegal) or medium (Thailand) in size. Their per capita income (expressed in U.S. dollars) places them high in the category of low-income countries (Senegal) or in the middle of the category of medium-income countries (Thailand and Costa Rica). All three countries are agricultural although the economies of two of them (Costa Rica and Thailand) have been structurally changing as industry accounts for an increasing part of the GNP. None of them numbered among the top 15 producers of mainstream science from the DCs in the beginning of the 1980s. Thailand was sixteenth, just after Singapore, and was seventh among the Asian countries (excluding Japan). Despite its small size, Costa Rica was eighth among the Latin American countries, just after Columbia and Cuba. Senegal and Ivory Coast shared seventh place among countries of sub-Saharan Africa (excluding South Africa), but this was largely thanks to input from expatriate French scientists working in Senegal (Gaillard and Waast 1988).

Industrializing Economies Still Heavily Dependent on Agriculture

With a population barely over 2.6 million in 1987 and a land area some ten times smaller than France, Costa Rica is one of the smallest and sparsely populated countries of Latin America. The population is composed of white descendants of Spanish settlers, a small minority of blacks (1.8%) who live mainly around the coast, and Indians (1%). With an official literacy rate of over 90% and life expectancy of 74 years, Costa Rica is one of the leaders of Latin America. Education and health have received special attention, primary education has been free of charge and compulsory for over a century, and the largest sector in the national budget is education (Costa Rica 1981).

Although Costa Rica relies essentially on agriculture, industry is generating a growing part of the GNP. In 1974, for the first time, industry contributed more than agriculture. In the early 1980s, agriculture contributed close to 17% (as against 44% in 1950), accounted for 70% of the national exports, and provided employment for close to 30% of the working population. Half of the agricultural output is exported; most export earnings come from coffee and bananas, then meat, sugar, and a certain number of lesser products (ISNAR 1981).

The main problems facing Costa Rica now are inflation, the energy crisis (the country spends 25% of its export earnings on importing oil and oil products), the sluggishness of the Central American Common Market, the drop in world prices for its exports, and its foreign public debt, which is close to U.S. $4 billion. This, proportionately, makes Costa Rica one of the most heavily indebted countries in the world. Again considering the size of the population, Costa Rica is one of the countries that receives the most aid; a large part comes from the United States and from international organizations such as the World Bank and the Inter-American Development Bank.

Senegal was populated in phases, which explains the ethnic mixture of the territory. The main ethnic groups are the Wolofs (approximately 40% of the population, mainly farmers and civil servants), the Serers (16%, farmers-herders), and the Peulhs (14%, herders-nomads). There are 40,000 Europeans (called *toubabs*) including 20,000 French (81% of them are in Dakar) and 10,000 Portuguese (mainly traders). Although the most important sector in the national budget is education (28.3% of the 1985–86 budget), the illiteracy rate was estimated at 77.5%. At birth, life expectancy is only 43 years. Most of the Senegalese are Sunnite Muslims (91%), and 30% of the married men are polygamous (Encyclopædia Universalis 1988).

In 1981, 81% of the working population was employed in the primary sector, which meant that Senegal was, by nature, largely agricultural although this sector only contributed 17.6% of the Gross Domestic Product (GDP) (World Bank 1986). The agricultural sector has undergone considerable change since 1960, with agriculture waning as livestock and fish production rose. Groundnut is still by far the most important industrial crop, but it too is on the decline. Senegal produces insufficient food crops, in particular cereal crops, and every year has to import large quantities of rice and wheat. Since the early 1980s, especially after the agricultural program was stopped, the production system deteriorated. The relation between the selling price of agricultural products and the cost price of factors of production is so unfavorable that agricultural inputs are being used ever less, and agricultural equipment is wasting away (Sène 1985).

The industrial sector only employed 6% of the working population in 1980 and 5% in 1985 (World Bank 1986). Between 1980 and 1986, the secondary sector generated annually between 29% and 33% of the GDP. In industry, the main sectors were textiles and processing plant oils, mainly (75%) in the Cap Vert Peninsula. Sources of energy are insufficient, and the main mining resource is phosphate, which is the fourth most important Senegalese export (Bonnefond 1987).

It is difficult to evaluate increases in the GDP. A fair estimate is a steady 1% per year (Bonnefond and Couty 1988). The economy has become sluggish because of stagnation in the rural sector, paralysis in the modern sector, and the drop in real government outlay since 1980. As the old-time factors of growth faded, compensation could have come—but didn't—from the fish industry, phosphates, and tourism. Results are especially poor in agriculture since the annual GDP growth rate between 1973 and 1984 was −0.2%. The per capita GNP dropped by an average annual 0.5% between 1965 and 1984 (World Bank 1986). The economic and financial situations are alarming. The trade balance is bright red because of the increase in oil prices and the decrease in groundnut and phosphate prices. Exports covered 81.5% of the imports in 1977 but only 55% in 1980 (World Bank 1986). The foreign debt is another burden for the Senegalese economy and is constantly growing. The outstanding debt was 901.2 billion CFA francs (US $1 = ± 250 CFA in 1990) at the end of 1985, or 75.9% of the GDP. The service of the debt, even after debt rescheduling, continues to rise; it reached 77.2

billion CFA francs in 1985, in other words, 31.7% of the value of Senegalese exports (Bonnefond and Couty 1988).

Thailand has about the same number of inhabitants (54 million) as France and occupies a land area (514,000 kilometers^2) that is also much the same. It ranks in the midway group of Southeast Asian countries. The majority of the people are Thai (79.5%, which includes 52.5% Siamese and 26.9% Laotians); there is a Chinese minority (12.1%) that lives mainly in Bangkok and controls most of the trade and industry. The other ethnic groups include Malays (3.7%) who live mainly in the south and Khmers (2.7%). The principal religion is Buddhism. The main expenditure in the national budget is servicing the debt (24.7%). Second place is shared by national defense and education, each receiving 18%.

Agriculture provides the livelihood for two-thirds of the population (four-fifths in 1960) and generates one-third of the GDP. After 1850, the substantial increase in the price of rice motivated the Thai farmers to concentrate on this crop with the result being that rice exports rose to become 25 times greater in the century between 1850 and 1950. The increase in rice production is the result of cultivating more lands, not higher yields; yields, on the contrary, have been going down (Muscat 1966). Thailand is now the world's leading rice exporter (3 million tons) and has given over half its arable lands to rice. Cassava and maize are being farmed on newly opened lands.

Although four-fifths of the population lives in rural areas, the rapid urban growth, especially in Bangkok, whose population rose from 1.7 million in 1960 to over 5 million in 1988, has created the very serious problem for the government of providing jobs for the growing urban population and also creating new jobs in the rural zones to stem the exodus to the capital. Since 1960 a shrinking agricultural sector has been contributing less to the GDP and offering less employment while the industrial and services sectors have been expanding. Industry has registered the highest production growth rate, while in terms of job opportunities, construction has been faring best. It was mainly in the 1960s and 1970s that the textile industry soared; it now represents 15% of the GDP and employs close to 300,000 people. The textile industry, along with American and Japanese automobile assembly plants, the Saraburi integrated steel plant, and the country's only cement factories are all concentrated around Bangkok.

It was also in the 1960s that the private sector attracted large numbers of foreign investors. In 1974 foreign investments represented close to 30% of the capital in this sector. Japan was the spearhead: it provided

42% of the foreign capital invested in this field in Thailand that year; the United States came second but had invested the equivalent of less than half the Japanese investment (Mingsarn 1981). The trade balance is negative. Most trade is directed to industrialized Western countries and Japan. International tourism offsets part of the deficit; but since the 1960s, and more specifically since the United States withdrew from Vietnam, economic development has been dependent on American aid.

Regional Disparities

These three countries have more or less pronounced regional differences. In Senegal, the little bit of industrial development in progress is concentrated on the Cap Vert Peninsula, where Dakar, the capital city, is located. In Thailand, national industry is largely concentrated in the highly productive central plain where most of the food crops, principally rice, are grown. This region, which includes Bangkok, the capital city, contributes over half of the GNP. The least-favored area is the densely populated mountainous north. The economic despair of the most underendowed regions is expressed through temporary and permanent migration to the cities, mainly the capital. Of the three countries, Costa Rica has the fewest internal regional disparities.

These disparities are also seen through the development of schools of higher learning and the establishment of scientific communities.

Young Scientific Communities

The history and development of scientific research in Costa Rica, Senegal, and Thailand first and foremost evidence the youth of the systems. All three have traversed serious growth pangs that were more or less controlled in the 1960s. These pangs were accompanied and followed by ruptures and dysfunctioning as the states, in the early 1980s, forsook their commitments, especially in Senegal and Thailand.

Modern science may have been born in Italy at the end of the sixteenth or the beginning of the seventeenth century, but it wasn't until the middle of the nineteenth century in Costa Rica and Thailand and a little later in Senegal that durable, tangible signs of its arrival could be seen. The first traces of research institutes and schools of higher learning started appearing at the end of the nineteenth century and became really visible during the first part of the twentieth century. And it wasn't until the 1960s and even more so the 1970s that national scientific communities started taking root and becoming bona fide institutions.

Penetration of Western Science and Colonial Heritage

There is no clear link between the inroads made by Western science and the colonial heritage during the 1960s and the countries' colonial past or absence thereof. But it is interesting that Senegal, which became independent in 1960 (Costa Rica in 1821), was far behind at independence time, especially in the field of higher education. Until 1967 most of the student body at the University of Dakar was French; the university did not start accepting large numbers of Senegalese until 1969. Whatever the case, the choice and strategies of the French colonizers in Senegal, the withdrawal of the Spanish colonizers in Costa Rica, and the determination to stay independent in Thailand had their effect on the mainstreams of higher education and research and still affect the development of national scientific communities.

As concerns Costa Rica, Spain expected very little from this poor colony, did next to nothing to contribute to its development, and, at the time of independence in 1821, left very few Costa Ricans equipped to fashion the country's future. The University of Santo Tomas, Costa Rica's first attempt to create a university, was opened in 1843, 22 years after independence. Since there was no higher education available in Costa Rica, the future members of the national government were sent to the University of San Ramon de Leon in Nicaragua to study law and public administration. Nicaragua played an important role in Costa Rican affairs and in educating a national elite on the morrow of independence. By the beginning of the twentieth century, the only branch of higher education that functioned normally in the country was the Faculty of Law, which explains why Costa Rica as a country and its academic world were long dominated by jurists. Starting in 1840, the University of San Carlos in Guatemala gradually replaced San Ramon de Leon as the training ground for young Costa Ricans studying law and medicine (Gonzalez 1976).

In Senegal, the French colonizers sought to organize technical services to make the lands of immediate benefit to France. Special attention was given from the outset to general agronomic research and especially to export crops. An experimental groundnut station was created at Bambey in 1921, in the heart of the groundnut basin. Agronomic research did not include a training component; it wasn't until 1979 that a specialized school of agronomy was created in Senegal. This largely explains why agriculture, which now accounts for close to half of Senegal's scientific potential, was controlled by French expatriate scientists for such a long time and actually, to a great extent, still is.

To farm these lands meant controlling the tropical diseases that threatened the French colonizers. The Institut Pasteur in Dakar was created in 1924 as the successor to the bacteriology laboratory created in Saint Louis in 1896. Contrary to the agricultural field, medical research included a training component; the Dakar School of Medicine, created in 1918, was the first block of the West African university system. Consequently, Senegalese scientists, as of the early 1960s, were much more active and much more self-reliant for training in the field of medicine (39% as against an average 22%) than in other fields.

Initiatives to create other institutes were taken slowly until after World War II. The Institut Francais d'Afrique Noire (IFAN), for instance, was officially created in 1936 but actually started functioning after World War II and expanding its staff in the 1950s (IFAN 1961). The center for soil studies (Centre de recherches pédologiques) was created at Dakar-Hann in 1949 and when it became part of ORSTOM-Dakar in 1960 grew to include new disciplines (Gleizes 1985).

In Thailand, the nineteenth century Thai kings adopted bits and pieces of Western knowledge, institutional patterns, and technologies as a way of guarding their independence against the French and British imperialism that was running rife in Southeast Asia (Muscat 1966). King Chulalongkorn (1868–1910) practiced an open-door policy to Western ideas, which motivated him to create a modern public service, schools of higher learning such as the school of medicine (1889) and the Ministry of Education (1892); in 1890 he even went as far as to send members of the royal family to study in Europe, mainly in England (MOSTE 1987).

The University and the Student Population

Costa Rica was the first of the three countries to establish a university. The University of Santo Tomas was created in 1843 but was closed in 1888 by the minister of education who felt that university education was a luxury his country could not afford and that it was more important to use resources available to develop secondary education. The university was finally reopened in 1940 and was called the University of Costa Rica. Thailand established its first university, the University of Chulalongkorn, in 1917; the first medical doctors graduated in 1930, and the first B.S. degrees were awarded in 1935. In Senegal, the University of Dakar only took shape in 1957.

Student populations stayed small, and relatively few diplomas were

Figure 6.1.   Students as a Percentage of Total Population, 1960-1985

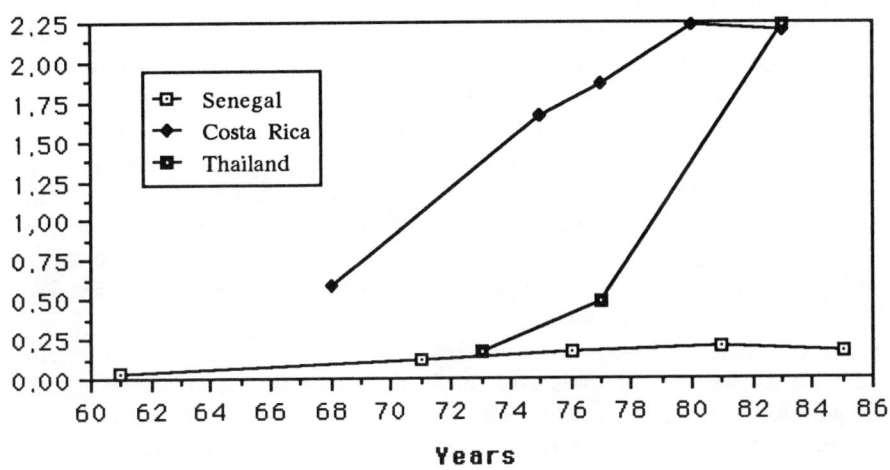

awarded until the end of the 1960s. In the 1970s student enrollment grew substantially in all three countries but at very different rates; in Costa Rica and Thailand during this period more establishments of higher learning were born. To facilitate a comparative understanding, the graph in figure 6.1 shows the student enrollment figures as a percentage of the total population in the years between 1960 and 1985.

By the end of the 1960s, Costa Rica had the highest student body (0.58% of the total population in 1968). Senegal and Thailand had percentages that were rather similar to each other (0.12% in Senegal in 1971 and 0.17% in Thailand in 1973). The student growth rate was fastest in Thailand toward the end of the 1970s; by the beginning of the 1980s, Thailand had caught up with Costa Rica. By 1983 the student population in these two countries amounted to 2.2% of the total population; this percentage figure is slightly higher than the equivalent French figure for that year.

On the other hand, in Senegal, notwithstanding the rise in enroll- ment from 2,500 to 10,000 in the 1970s, the student population still only amounted to 0.2% of the total population by the early 1980s, in other words ten times less than in Costa Rica and in Thailand. Part of the explanation for this difference lies in the new higher level schools that were built in Costa Rica (4 new universities were established between 1973 and 1978) and in Thailand (11 of the 14 public universities were created after 1960). In Thailand the proliferation and growth of private universities, mainly specializing in business, also contributed to this spectacular development, while in Senegal, the University of Dakar, even now, is the only Senegalese university, since the plan to build a university at Saint Louis failed.

This rapid increase in both public and private universities in both Thailand and Costa Rica, of course, has had an effect on teacher training and on research methods. In the beginning, the new universities had to draw on part of the teaching staff from the older universities and, to a small degree, on foreign staff. Thereafter, the teaching staff was expanded from within by hiring new graduates. As a result, the teaching staff in the newly created universities was very young in both age and educational experience. Nevertheless, the majority of the scientists and the best trained researchers work in the universities. We will come back to this later.

The student boom and the large number of graduates, combined with the economic crisis and the budgetary cuts, gave rise to a new phenomenon, especially in Senegal and in Costa Rica, viz., unemployment among the intellectuals. In Costa Rica, the "production" of 4,000 new graduates each year since 1980 has led to relative saturation of the labor market in various professions, especially in law, medicine, and senior positions in agriculture. This phenomenon has reached an alarming level in Senegal where nearly all branches suffer. An association of unemployed university graduates, organized per discipline, was born in 1981. This does not mean that all the employment needs have been fulfilled. The situation is quite the opposite. But the key employer, i.e., the state, is no longer able to keep up with the need to create new posts.

As universities grew, research activities went through an institutionalization process, and science policy-making bodies were created. This process started in Thailand at the end of the 1950s, in Senegal in the 1960s, and in Costa Rica at the beginning of the 1970s.

Creation of Science Policy-Making Bodies

With an eye on the European models, the three countries in our study built up organizations to handle scientific policy. As institutional structures developed, so did these organizations; but in the end, all three countries put science and technology under a ministry of its own: the Ministry of Science, Technology, and Energy (MOSTE) was created in Thailand in 1979; the Ministry of Scientific and Technical Research (MRST) was created in Senegal in 1983; and the Ministry of Science and Technology (MYCIT) was created in Costa Rica in 1986.

The MRST did not live very long. During a ministerial reorganization in 1986, it was replaced by the Department for Scientific and

Technical Affairs (DAST), created as part of the Ministry of Planning and Cooperation. Unlike MRST, DAST has no financial or managerial responsibility for national research activities. Although not of ministerial rank, DAST works much like MOSTE in Thailand and MYCIT in Costa Rica, which, because of the institutional organization of science and technology, do not have custody of or managerial responsibility for research organizations (except in one case in Thailand). Actually, the research institutes fall under various ministries, depending on their field of activity, i.e., agriculture, health, industry, education, etc.

Ministries for science and technology were grafted onto an institutional fabric without replacing any of the threads. The existing organizations were brought under their responsibility with varying degrees of success. The systems caused many problems especially because of bureaucracy and power struggles. The first problem was the growth of bureaucracy, especially painful in a country with a small scientific community like Costa Rica. The creation of a ministry generated a big need for qualified staff without decreasing the staff requirements of ancillary institutions responsible for coordinating and promoting research activities, such as the national research councils (NRC in Thailand and CONICIT in Costa Rica). All these institutions, especially within the universities, called on outside consultants for various councils, commissions, and work groups, thereby depriving the best scientists of precious times for research. The second major problem was related to the jurisdiction of scientific research management organizations. Superimposition of these organizations led to power struggles and the need for supracoordination, e.g., Thailand where there was a need felt for an interministerial coordination body.

These bureaucratic organizations, especially in Latin America, are also often accused of spending over half their budget on internal operating expenses (Moravcsik 1979). Furthermore, despite the political decision underlying their creation, institution budgets are too small to run their programs, which, therefore, have to rely on outside loans and grants, e.g., CONICIT in Costa Rica. Further, there is general agreement that investments in scientific activities need to be coordinated at the highest level, but interference by science policy-making bodies can be dangerously excessive. University scientists are constantly being accused of running luxury research projects that the country cannot afford, even when the amounts involved are negligible. This situation is difficult to accept for scientists who are witness to the creation of expensive, generally inefficient scientific bureaucracy and the imposition from above of often simplistically ranked priorities. Of course, a minimum of planning and perspective is essential. But

Table 6.2. Estimated Number of Scientists

Country	Total No. of Scientists	No. of Scientists FTE	No. of Scientists FTE Per 1,000 Inhab.
Costa Rica[a]	800–1,000	400–500	0.17
Senegal[b]	1,000–1,100	500–600	0.08
Thailand[c]	5,000–5,500	3,000–3,500	0.06

a = 1981
b = 1984–85
c = 1982
FTE = full-time equivalents

thought has to be given to the size of the system to be run and the need to avoid establishing a management system that ends up by serving its own interests rather than helping to structure a budding scientific community.

A Scientific Potential Difficult to Evaluate

Let us not forget that we are dealing with small scientific communities that were estimated, in the early 1980s, at 800 to 1,000 people in Costa Rica (CONICIT/IDRC 1982) and in Senegal (Gaillard 1989) and just over 5,000 in Thailand (MOSTE 1987). It is difficult to obtain confirmed figures on the numbers of participating scientists and their breakdown per institution and per major field of science. Most of the surveys available are incomplete or out of date. It is also difficult to make comparisons between countries because of differences in data and definitions. The sources we consulted, however, provide general ideas of scope and trends (table 6.2).

Any evaluation of the national stock of research scientists in full-time equivalents (FTE) is strongly affected by the concentration of large numbers of scientists in the field of higher education. University scientists only devote part of their time (one-third, according to our estimates) to research. In an effort to assess the importance of research work relative to the national population, we calculated the number of FTE scientists per 1,000 inhabitants. Costa Rica, the least populated country, has the highest figure (0.17), with two to three times more than Senegal or Thailand. Costa Rica places midway between Nigeria (0.06) and Brazil (0.37) in the league of DCs but far behind the most advanced ICs. There are approximately 20 times fewer FTE researchers per 1,000 inhabitants in Costa Rica than in France and 30 times fewer

Figure 6.2. Distribution of FTE Scientists within Institutions

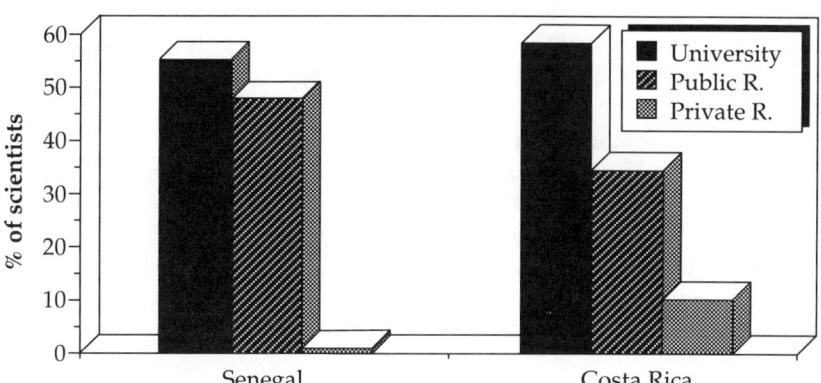

than in the United States. Further, simple proportionality cannot be used as the general rule because in the event of problems of scale it is disadvantageous for the small countries.

In sum, the number of FTE research scientists as a percentage of the population as a whole is very small, compared to the ICs. It is interesting to see the breakdown between types of institutions and fields of science.

More than half of the FTE scientific potential is to be found in the universities (56% in Senegal, 58% in Costa Rica) (fig. 6.2). We do not have global statistics on the assignment of Thai scientists per type of institution; a variety of indices points toward the higher education sector to find the vast majority of them. As an example, we see that 80.7% of the abstracts from the 1984 annual symposium of the Science Society of Thailand are signed by university scientists (Yuthavong 1986).

University scientists are the best trained and have the most degrees. Their average level of education rose considerably during the 1970s. In Senegal, for instance, university scientists have more degrees and are older than their colleagues in research institutes. Using the 1981 AUPELF (Association des Universités Partiellement ou Entièrement de Langue Française) file (AUPELF 1984), we saw that over half (55%) had finished their doctorate (third-cycle) studies. In Thailand, some four-fifths of the teachers/scientists at least have their M.S. degree, and most of them graduated from Thai universities. Close to one-third have a Ph.D., in most cases conferred by an English-language university abroad (Gaillard 1989). It is difficult to make precise comparisons between various avenues of education, but very probably the Costa Rican teachers/researchers have the lowest level of training. A report from the University of Costa Rica (UCR 1985c) indicated that only 20% of them had their doctorate and that the majority (60%) only had their bachelor of science.

The teaching staff of the young universities is younger and less well trained, in particular in Costa Rica. Out of an academic staff of 1,229 at the national university (UNA) in 1985, only 79, i.e., 6.4%, had a doctorate. At the Costa Rica institute of technology (ITCR) in 1984, barely more than 1% of the teaching staff had a doctorate. But then, the staff was very young: 48% was under 30 years of age, and 80% was under 35. These statistical figures reflect and imply problems inherent in the creation of several universities within a few years' time when there is a shortage of qualified teachers. By way of illustration, UNA and ITCR were only able to start their activities by drawing on the teaching potential from UCR and by bringing in—avowedly, in very small numbers—foreign staff. Afterward, the teaching staff was expanded from within by recruiting young graduates, which explains the relatively low level of education, especially at ITCR. To improve the situation, some but not enough scholarships are awarded for staff to study abroad, in particular to reach the Ph.D. level.

University scientists in general are better trained, but they are often the poor cousin of research in their home countries. This is especially true in Senegal where, according to our calculations, university FTE scientists have budgets that are 5 to 10 times lower than those of their colleagues in research institutes. A scan of the various institutes shows that research at the Senegalese institute for agronomic research (ISRA) consumes a very large part of the Senegalese national research budget. Similarly, of the 13 ministries that participate in R&D activities in Thailand, the Ministry of Agriculture stands out clearly since it consumed 54.5% of the national R&D budget in 1985 while the Ministry of Universities only consumed 7% that same year (MOSTE 1987). This explains why the Thai universities often have to find their own resources or appeal for outside, national, or foreign resources to round out their research budgets.

The situation is less clear-cut in Costa Rica where, at least in the early 1980s, the universities consumed about half of the national research budget (CONICIT/IDRC 1982). Whatever the case, many of the research programs conducted in the universities would not have been possible without foreign aid. University scientists, of a fairly good

Figure 6.3. Distribution of FTE Scientists in Major Fields of Science

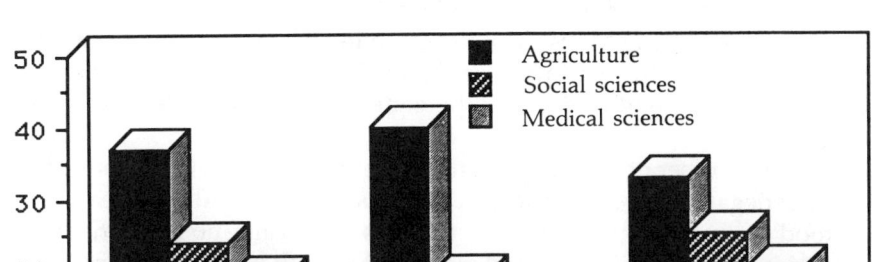

level, constitute a potentially important source that is often under-
utilized.

The distribution within the different types of institutions also illustrates the weakness of private research in Costa Rica and in Thailand
and its virtual nonexistence in Senegal (1%). The percentage figure is
slightly higher in Thailand but, in numbers, only involves somewhere
between 100 and 200 scientists. A recent study has also shown that the
demand for highly qualified scientific personnel in the Thai private
sector is very limited. Costa Rica has a higher demand, but the private
sector seems exclusively interested in social science research and, to a
much lesser degree, agriculture. This is a situation found in many Latin
American countries where social science research is gradually moving
out of the universities and into research centers or, even more so, into
private consultancy offices. In Costa Rica, thus, over one-third (36.5%)
of social science research is conducted in private centers, although the
university still has the leading role. A clear understanding of the gap
between Costa Rica, Thailand, and Senegal on the one hand and the
ICs on the other can be found in a figure for the OECD countries,
where, on the whole, 50% of the research budget is used by the private
sector, mainly for industrial research (60% in some NICs of Asia).

A breakdown of scientists available in the main fields of research
(fig. 6.3) shows concentration in three major fields: agriculture, social
sciences, and health. Agriculture is by far the leader in Thailand (33%
of the scientists) and Senegal (40%). The social sciences are in second

place in all three countries (about 20%). Health is clearly third in Costa Rica and in Thailand (20%), while in Senegal, according to a survey carried out by the Ministry of Research in 1981, it involved only 4.5% (MRST 1981). This percentage is obviously underestimated,[2] but in any case, the figure is well below the equivalent figure for the two other countries.

The number of scientists in engineering is low in all three countries. In Senegal barely more than 5% of the scientists conduct research on food technology. Industrial research is practically nonexistent there and is very minor in Thailand and Costa Rica. It involves a mere 6–7% of the research scientists, which is far below the needs of these countries where industry is rapidly growing. The scanty results of the incentive programs Costa Rica introduced to make up for the lack of effort in this field are indicative of the shortage of scientists qualified to work on industrial research and engineering.

On the other hand, the importance of the number of research scientists in Senegal working on what we could call "general advancement of knowledge" is rather unusual and can only be explained by the fact that ORSTOM has a large contingent of French scientists working there. This leads us to one of the main differences between the three scientific communities. The Senegalese scientific community, unlike the other two, as was mentioned above, is still largely dependent on expatriate scientists.

Degree of Dependence on Expatriate Scientists

The analytical report of potential scientific and technical resources in Senegal in 1975 by Gillet (1976) has a chapter on human resources that draws a robot picture of the "average" Senegalese research scientist 15 years after independence: the person is 39.1 years old; there is a 91% chance that the person is a man, a 58% chance that the person is French, and a 30% chance that the person is Senegalese!

Then comes the question of time allotted to various activities, years of experience in research, publication rate, and conference attendance. Early figures are especially indicative of the heavy reliance of the Senegalese scientific community on French scientists. In the 1973 survey, out of a total scientific community of 416 scientists, only 20% were Senegalese (CNPRS 1973). Senegalese scientists, furthermore, tended to concentrate in specific disciplines such as medicine (39%, as against an overall average of 22%) and social sciences (18%, as against 13%), while there were far fewer than the expatriates, mostly French, in agronomy and biology (CNPRS 1973). The non-French expatriates

Table 6.3. Breakdown of Senegalese and Foreign Personnel in Main Institutes of Higher Education and Research in Senegal, 1984–1988

Institute	Expatriates		Senegalese		
	No.	%	No.	%	Total
University[a]	191	40.0	287	60.0	478
Institutes affiliated with university[a]	107	47.0	120	53.0	227
ISRA[b]	92	47.0	104	53.0	196
ITA[b]	2	7.0	27	93.0	29
ORSTOM[c] includ. senior technicians and engineers	109	88.0	15	12.0	124
Institut Pasteur[c]	17	71.0	7	29.0	24
Subtotal	518	48.0	560	52.0	1,078
Agric. eng. and same category, rural dev.[d]	46	41.0	67	59.0	113
Average	564	47.0	627	53.0	1,191

[a]1984 staff
[b]1985 staff
[c]1988 staff
[d]1980 staff

came mainly from neighboring French-speaking countries and worked essentially in the university. By 1973 there were a few Belgian and Dutch scientists, and of late there have also been some Americans and Asians.

Senegalization of the scientific community gained such momentum in the 1970s that now the expatriate/Senegalese ratio has practically been reversed. A survey started in 1981 on 828 scientists pointed to this inversion; 75% of the scientists were Senegalese, 20% were French, the remaining 5% included mainly non-Senegalese Africans, but there were also 9 Belgians, 3 Italians, 2 Englishmen, 2 Americans, and 1 Indian. The 1983–85 figures we have on teachers/scientists working in the university and the main research institutes, viz., ISRA (Senegalese institute for agricultural research), ITA (Food and nutrition research institute), ORSTOM (French institute of scientific research for develop-ment in cooperation), and Institut Pasteur, show 52% to be Senegalese (table 6.3). The total percentage of Senegalese is practically the same with or without agricultural engineers and others in the same category working on rural development.

With the exception of ORSTOM, ITA, and Institut Pasteur, the per-

Figure 6.4.   Changes in Numbers of National and Expatriate Scientists at ISRA, 1983-1988

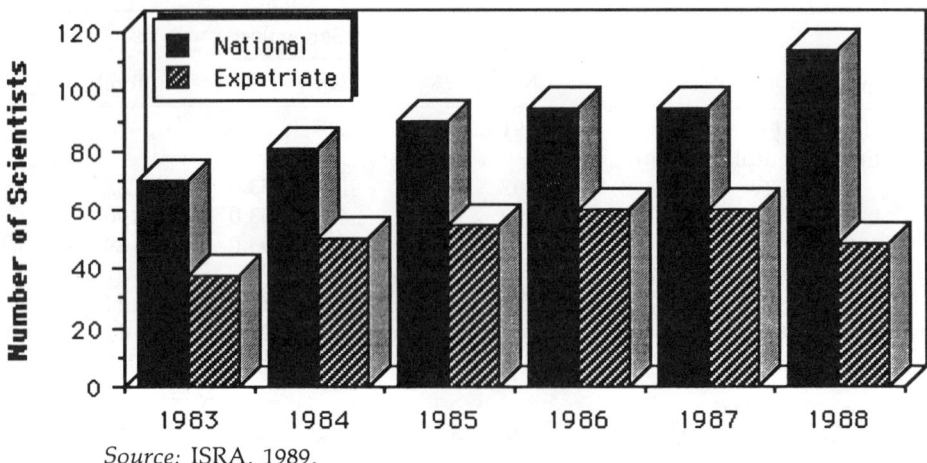

Source: ISRA, 1989.

centage of Senegalese was somewhere between 53% at ISRA in 1984 and 60% at the university in 1983–84. In this example also the university has the highest number of non-Senegalese Africans. Including this figure, the Africanization rate for teachers/researchers at the university in 1983–84 was 66%. ITA is a Senegalese innovation created after independence. France participated very little, if at all, which explains the low percentage of resident foreign scientists. ORSTOM and Institut Pasteur are exceptions, since they are the last French public services with a non-Senegalized research center operating in Senegal, hence the heavy presence of French scientists. Efforts have been made, especially at ORSTOM, to include Senegalese staff in programs designed together with Senegalese research organizations.[3]

The percentage of Senegalese in the national scientific community has increased sharply since the beginning of the 1970s when they only accounted for 14% of the scientists working in agriculture, for instance. It is encouraging to note that now the majority of the ISRA scientific staff is Senegalese (fig. 6.4).[4]

Unlike Senegal, Costa Rica and Thailand hire very few foreign scientists; there were 156 (including 87 Japanese, 33 Americans, and 28 Europeans) in Thailand in 1986, i.e., about 3% of the national community (NRC/MOSTE 1987), of which two-thirds worked on human sciences. Foreign scientists are proportionately more numerous outside the capital city than the Thai scientists, because they often work with foreign aid programs that are increasingly being sited in the more

Table 6.4. Distribution of Teachers and Scientists at Chiang Mai
University, by Sex and Academic Level, 1986

| | Men | | Women | | |
	No.	%	No.	%	Total
Lecturer	346	45.5	414	54.5	760
Assistant professor	260	50.0	260	50.0	520
Associate professor	112	73.0	42	27.0	154
Professor	15	88.0	2	12.0	17
Total	733	50.5	718	49.5	1,451

Source: Chiang Mai University Information Book, 1986, table 4, p. 24.

remote areas. Furthermore, there are not many foreign teachers in the
Thai universities; out of a teaching staff of 2,401 at the Chulalongkorn
University in 1985, 61 were not Thai and had come mainly to teach
foreign languages (CUB 1988).

Participation of Women in Research and Higher Education

Disparities between the three countries are also very pronounced as
concerns the participation of women in research and higher education.
Thailand has the most and Senegal the least.

In Thai universities, there are almost as many women teachers as
men. In Chiang Mai University in 1986, 49.5% of the teachers were
women. There seems to be a clear dichotomy in discipline preferences
between men and women; women are more strongly attracted to
medical and paramedical professions and to the social and human
sciences than men and seem to work at lower academic levels (table
6.4).

In Thailand women are strongly represented not only in the universities but also in public research institutes such as the Thai Department
of Agriculture (DoA). A recent study indicated that out of the 1,419
scientists at the DoA 38% were women (Elliot 1984). As far as we know,
this is the world record, especially since agricultural research institutes
have traditionally had very few women on their research staff.[5] Furthermore, most of the women are young; half of them are under 35
years of age. There are, moreover, more women (269) than men (259) in
this age group. Considering recruitment trends and retirement figures,
the percentage of women in the research services of the DoA is expected to exceed 50% by 1994. Although we do not have precise

Table 6.5. Distribution of Teachers and Scientists at UCR by Sex

	Teachers		Scientists		Total	
	No.	%	No.	%	No.	%
Women	518	26.7	134	31.0	652	27.4
Men	1,429	73.3	298	69.0	1,727	72.6
Total	1,947	100.0	432	100.0	2,379	100.0

Source: UCR 1985.

statistics on the other research sectors, we have seen that the same trend applies to other sectors, except engineering.

The medium- and long-term implications for research applications and orientations in Thailand are very important. There are very marked differences between men and women in the choice of disciplines, mobility, power relations, etc. These implications show up clearly in the study on DoA personnel. A key issue is the place of work. Only 29% of the total research staff work in centers located outside of Bangkok, which, for an agricultural research institute, is very little. The breakdown is 20% women and 34% men. Since the percentage of women working for DoA is predicted to continue rising, the department can expect to have more and more difficulty in assigning staff to work outside of Bangkok.

In Costa Rica (table 6.5) women account for between one-fourth and one-third of the teacher/scientist staff at the public universities, e.g., 27.4% at UCR. Paradoxically, percentagewise there are only half as many women as men who have their Ph.D.s, and 31% of them work on research (UCR 1985b and c).

The choice of discipline is also strongly related to sex. Women rank their preferences as follows: arts (47%), social sciences (36%), health (21%), basic science (20%), engineering and architecture (7.5%).

In Senegal women constitute a minority both in the university and in the public research institutes. At the University of Dakar, only 11.7% of the teachers are women (among the Senegalese teachers, only 9% are women). Women are best represented at the Faculty of Medicine and Pharmaceutical Studies (16%) and least represented at the Faculty of Law and Economics (4%). There are proportionately far fewer in the higher academic echelons such as senior lecturer (3%) than at the lower levels such as assistants (26%).

Table 6.6. Place of Education of Senior Senegalese Scientific and
Technical Staff Working at ISRA, 1985

Country	No.	%
1. France	56	43
2. United States	28	21
3. Western Europe (excl. France)	15	12
4. Eastern Europe (incl. USSR)	12	9
5. Senegal	12	9
6. Canada	5	4
7. Other	3	2
Total	131	100

Source: Sène 1985, p. 31.

Reliance on Research Training Abroad

Paradoxically, the countries with the youngest university systems such
as Senegal offer the most complete and elaborate doctoral level curricula. But the University of Dakar does not have enough qualified
teachers and scientists to supervise Ph.D. (third-cycle) students. The
possibility of studying at that level and then building up a career in
research also depends on the discipline. This explains why a large
number of doctors and veterinarians are trained in Senegal. In other
fields, such as agricultural sciences, training still often takes the student abroad. Out of a scientific and technical staff of 131 senior Senegalese who were working at ISRA in 1985, only 12 (9%) had completed
their advanced studies in Senegal (table 6.6). This is relevant to understanding why it took until 1979 for a specialized school of agriculture to
be built in Senegal. It may a priori seem paradoxical and abnormal for a
country that spends 50% of its research budget on agriculture and has
considerable research strength in this field not to have drawn earlier on
its experienced scientists to train its research staff.

A look at the list of countries where the ISRA scientists were trained
shows that because of the diversification in fields of specialization the
preference was to study in the United States, or in Western (not only
France) and Eastern Europe. France is still the main source of education
for Senegalese scientists working in nearly all fields, having trained
43% of the ISRA staff. The United States has only been working in
Senegal for a few years and seems to be specializing in a limited
number of disciplines, in particular in rural sociology and economics. It
is expected to make greater efforts in the coming years.

In Thailand, despite the fact that the three main universities

Table 6.7. Place of Education and Educational Levels of Science and Technology Teachers at Chulalongkorn University

Country or University	B.S.		M.S.		Ph.D.	
	No.	%	No.	%	No.	%
Chulalongkorn	110	84.0	78	52.5	5	3.5
Other universities	11	8.5	20	13.5	9	6.0
Total Thailand	121	92.5	98	66.0	14	9.5
United States	10	7.5	50	34.0	83	56.0
Europe	—	—	—	—	51	34.5
Total outside Thailand	10	7.5	50	34.0	134	90.5

Source: Graduate School Announcement, 1986–87, Chulalongkorn University.

(Chulalongkorn, Mahidol, and Kasetsart) set up graduate schools in the 1960s, there are very few Ph.D. training programs, as the following examples show. The Chulalongkorn University in academic year 1984–85 had a graduating class of 3,015 students, of which 818 received their M.S., but only 3 graduated with a Ph.D. In 1986 the Kasetsart graduate school enrolled 2,163 students in their M.S. program but only 40 for the Ph.D. The number of Ph.D. candidates reached 55 in 1987 in the (only) five disciplines offering a Ph.D. program, viz., 6 in entomology, 5 in plant pathology, 32 in soil sciences, 10 in agronomy, and 2 in horticulture. There are several reasons for this situation, e.g., many of the Ph.D. candidates have a job while preparing their thesis and tend to take an inordinately long time in writing it. Another reason is that thesis supervisors are difficult to contact because of their many commitments outside the university. The number of Ph.D. graduates in Thailand is far too small. There are not even enough to replace the professors who retire. The situation is especially alarming since scholarships to study abroad are becoming much scarcer.

Most of these professors had scholarships to complete their university studies abroad some time between the middle of the 1950s and the end of the 1970s. Very few studies to the Ph.D. level within the country, as we can see from a quick survey of the academic background of 148 Chulalongkorn University professors affiliated with the graduate school program on science and technology. Table 6.7 shows that 92.5% studied for their B.S., 66% their M.S., and 9.5% their Ph.D. in Thailand. Out of the 14 Ph.D.s awarded in Thailand, 8 were received in microbiology and biochemistry from Mahidol University. It is also interesting that the great majority of the Chulalongkorn University teachers in our sample went to school at Chulalongkorn University.

Now there are far fewer young Thais studying abroad. According to UNESCO, in the early 1980s there were close to 10,000 in foreign universities. Over half (58%) studied in the United States (UNESCO 1985a), and 20% studied in two Asian countries, Philippines and India, which thus ranked second and third. In 1983 Japan ranked fourth by enrolling 378 Thai students. Then came European countries, led by France and Federal Republic of Germany.

In Costa Rica there is no Ph.D. program; the highest degree awarded by a Costa Rican university is the "maestria." There are 32 "maestria" programs; over half are offered at UCR. Furthermore, since the beginning of the 1980s, three-fourths of the "maestria" have been in public administration (21) and agricultural sciences and natural resources (22). Except in agricultural sciences, there are very few "maestria" programs created since 1975 that lead to a career in research. Costa Rican students who want to study for a Ph.D. therefore are forced to go abroad.

According to UNESCO statistics, in 1981 there were 969 Costa Rican Ph.D. candidates studying abroad; over half (57.68%) were in the United States, 20.5% in Europe, and interestingly enough 11.45% in Latin America (Argentina, Guatemala, and Cuba). Some students benefited from programs administered either by their universities or by CONICIT, usually sponsored through bilateral agreements; and some students left on their own. Between 1973 and 1982, 127 Costa Rican students were able to study for an M.S. or a Ph.D. in the United States under a U.S. AID supported program administered by CONICIT. UCR runs the biggest Ph.D. scholarship program; between 1971 and 1984, it administered 549 (Gonzalez and Calderon Alfaro 1986). The number of scholarships fluctuates from year to year but tended to go down during the last few years of the reference period. Between 1971 and 1977, 50 scholarships were awarded every year, while between 1981 and 1984, the annual figure dropped to under 20. This drop was largely due to the devaluation of the national currency, the colon, in relation to the U.S. dollar, as of the early 1980s.

Close to half (42.3%) of the UCR students who received scholarships went to study in the United States. The position of the other countries was much similar to data UNESCO reported for 1981. The spearhead position of the United States is due to the prestige of U.S. universities, the relative ease in learning English, geographical proximity, the large number of scholarship programs available to UCR, and the sophistication and large number of scientific and technical courses. The last reason explains the choice of many Costa Ricans to study agronomy, engineering, physics, biology, microbiology, computer sciences, and economics in the United States. The 13.8% of the students who use their scholarships in France, the second country on the list, prefer disciplines such as sociology, law, journalism, political sciences, in other words, the social sciences; but they are also interested in the arts and medicine. Third on the list is England where 36 UCR students got scholarships; during the first few years of the reference period, England played an important role, but slowly but surely scholarships became scarcer. The possibility of getting a scholarship from a country plays an important or even a decisive role. Gradually various countries seem to specialize in selected disciplines. It was also interesting that the three countries that offered the most scholarships required the Costa Rican students to learn a foreign language, which apparently was not a problem.

In sum, the United States was the first choice for the Thai and Costa Rican students, and France was the first choice for the Senegalese students. As of the early 1980s, subject specialization started becoming so important that considerable numbers of Senegalese, especially those working on agricultural sciences, decided to study in the United States; and, similarly, in more recent years, 76 Costa Ricans (out of 549) from UCR decided to use their scholarships to study in France. Some of the DCs are facilitating enrollment for foreign students also, e.g., Thai students in India and Philippines; Costa Rican students in Argentina, Mexico, and Brazil; and, to a lesser extent, Senegalese students in Ivory Coast and Morocco.

We feel that research training is too heavily reliant on foreign facilities and that training abroad does not satisfy the quantitative or qualitative needs of the scientists from the three countries in this report. We feel that it would prove to be more realistic, efficient, and, in time, productive to allocate the considerable sums of money[6] now being used to train future scientists abroad to establish doctoral (third-cycle) programs leading to a Ph.D. in priority fields within the national universities. University training within the home country would also contribute to structuring the scientific "fabric" and strengthening the national scientific communities. This implies that the countries in the northern hemisphere would have to remodel their education aid policy but obviously does not mean eliminating the possibility for postdoctoral education abroad in certain very highly specialized or marginal fields.

Scientific Production Not Very Much, Not Very Visible

Even in relation to their home continents, the three countries in this report are small science producers; Senegal and Costa Rica produce between 50 and 100 mainstream publications a year, Thailand about 300. Costa Rica's output, thus, accounted for about 1% of the aggregate Latin American production. Thailand hardly scored better: 2% of the Asian output (excluding Japan). Senegal produced about 4% of the sub-Saharan African scientific output. As a percentage of total production, these countries don't "weigh" much compared to Third World giants such as India, Brazil, and Nigeria. The per scientist production, calculated in numbers of mainstream publications per scientist per year, is small: about one-sixth in Costa Rica and one-tenth in Thailand and in Senegal. This varies according to discipline and institution.

Authors of mainstream publications often work in universities. This is the case for over half in Senegal and over two-thirds in Costa Rica and in Thailand. In all three countries, the most "productive" and "visible" disciplines are health, biology (especially in Costa Rica), and environmental sciences (especially in Senegal). Social sciences are not very "visible," except in Costa Rica where they account for close to one-fifth of the total output. Agricultural sciences, which have the greatest potential for the future in all three countries, are hardly "visible" even when they are productive. An analysis of the total scientific production of ISRA scientists in Senegal showed that the vast majority of the scientists did write but that their research data did not very often get published in specialized journals.

Actually, research findings and reports produced by research scientists in these three countries are often only stenciled and circulated within the institute, if they are circulated at all. The publication strategy for scientists in these three countries depends on their institute's editorial policy and the number of local journals in existence. Of the three countries in this report, Thailand has the most local journals (around 200) and the most scientific output. There is also a close correlation between the number of articles the scientist produces for publication in local and in international journals. In other words, the most productive scientists publish just as much in the local reviews as they do in the international journals. Costa Rica and Senegal have a much smaller number of local journals. In Costa Rica journals appear rather regularly while in Senegal they do not. Consequently, the small number of journals that are published rather regularly are so heavily solicited that scientists are forced to submit their research for publication in journals abroad.

This is exactly the case for scientists in the three countries, especially in Costa Rica and in Senegal. Actually, the percentage of articles published within Senegal has dropped from about 30% in 1975 to 23% in 1985, while during that same decade the percentage of Senegalese scientists in the national scientific community increased substantially. In Senegal, like in most French-speaking countries of Africa (but not in Costa Rica or in Thailand), over two-thirds (68.5%) of the authors were foreigners, mainly French. The productivity rate of Senegalese and Thai, and to a certain extent Costa Rican, scientists is low because of professional status, working conditions, lack of incentive, etc.

Research Scientists in Search of Statutes and Status

Of the three countries, Costa Rica offers its scientists, especially those working at UCR, the best salaries and the most incentives to delve into scientific activities and publish their findings. Proposals have recently been made through the UCR internal promotion system to encourage teachers/scientists to do more research. Obviously, there is still room for improvement, and different institutes offer different opportunities. The big devaluation of the colon early in the 1980s seriously affected salaries and lowered the purchasing power. But this situation was not unique to people working in science.

In Thailand, despite a political decision to support scientific work and an atmosphere propitious to its research development, there are inherent difficulties in the profession that are far from being solved. In the public sector, Thai research scientists and university teachers are badly underpaid. Promotions and salary increases depend almost exclusively on seniority, with very little credit for educational level, work performance, and services rendered, be it in research, education, or administration. This situation is, of course, deleterious to normal research schedules since far too many research scientists and teachers doing research have to round out their salaries by accepting unrelated side jobs.

The status of the research scientist in Senegal is precarious. Since there is no career stability, a future in this profession is fraught with uncertainty. Institute regulations make no allowances for the unique characteristics of the profession. Career paths and positions are very different from each other. The scientists who are assimilated to civil servants are probably worst off since the professional scale in the government civil service does not provide for a Ph.D. level. This means that scientists with a doctorate, a master's, or a bachelor's degree are all in the same category and move up the seniority-driven scale at the same pace. Attempts to provide the research scientists with common professional statutes were nearing completion when the MRST was dissolved. Considering the present institutional structure, the eco-

Figure 6.5.   Research Budget as a Percentage of GNP in Costa Rica, Senegal, and Thailand, 1978-1986

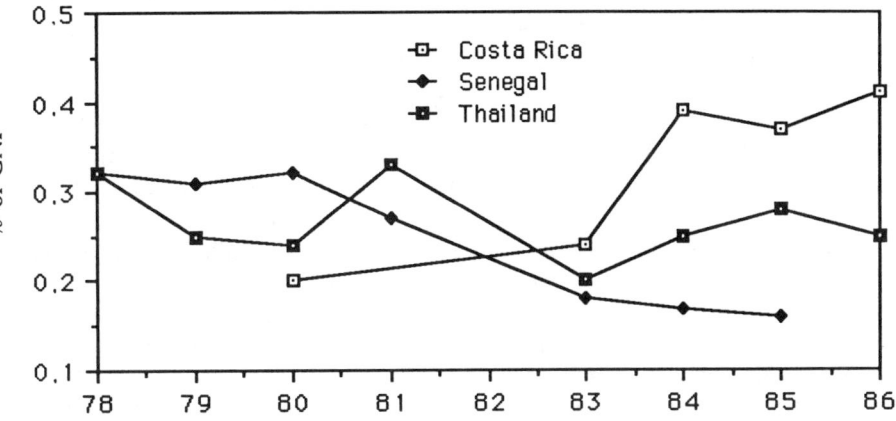

nomic crisis, and the budgetary restrictions in Senegal, there is little chance that research will be given statutes of its own in the near future.

Insufficient Overall Financing

As concerns research financing, the three countries are far from their targeted goals. Costa Rica is the only country that has made a noteworthy effort to increase the national research budget since the beginning of the 1980s and in 1986 devoted 0.41% of its GNP to research (fig. 6.5).

Thailand, and even more so Senegal, has invested less in research since the beginning of the 1980s and, expressed as a percentage of the GNP, now invests 10 to 20 times less on research than the OECD countries. The aggregate value of this indicator hides large intersectoral differences. For the three countries, thus, we see that the biggest effort is made in agriculture. The relatively smaller financial effort that has been observed in Senegal and Thailand will mean stronger reliance on foreign sources of funding and imbalances in research activities. Of the three countries, Senegal relies most heavily on foreign resources that now fund two-thirds of its research budget. Foreign sources of funding have been diversified during the last decade although most aid for Costa Rica and Thailand comes essentially from the United States, while Senegal turns to France. This said, French financing of Senegalese research has gone down from 59.3% in 1972 to 35.5% in 1986. By way of compensation Senegal has been able to obtain support from the United States (mainly through U.S. AID) and the World Bank. These two sources alone have met close to one-fourth of the Senegalese research costs in 1986, while earlier, in 1975 for instance, they provided very minimal support. Foreign financing of Thai research is less important, although in certain years it accounts for 30% of the total outlay. We do not have global statistics for Costa Rica but know that its reliance on foreign sources to fund research is much the same as in Thailand.

Foreign financing is especially important in agriculture and health. Incentives and new, well-tailored programs have recently been set up with foreign aid to offset personnel shortages in engineering and in industrial research, mainly in Costa Rica and in Thailand. Interest in these new programs has been slow because of the shortage of qualified scientists. The private sector finances very little research, a mere 1% of the total budget. Here again, a comparison with the OECD countries, where the private sector supports at least half of the research costs, reveals extreme imbalances in research funding systems in the DCs.

The research budget in the universities, where more than half of the research expertise is to be found, is especially small. The University of Dakar in Senegal does not have its own research budget, and it is difficult to calculate how much money is actually spent on research. According to a UNESCO/UNDP (United Nations Development Program) study, research costs represented 3.6% of the total university budget in 1981. The example of health-related research in Senegal enabled us to understand the difference in status between university and nonuniversity research. In Costa Rica the UCR research budget has increased substantially during the last few years. In 1984 it represented close to 12% of the total UCR budget. But it is still far too small; and all the universities have to look for outside, especially foreign, funding. This also applies to Thai universities. We have seen that the Chulalongkorn University had to count on foreign financing for more than 50% of its research budget. The resources made available to university scientists are so pitifully small that foreign funding has become decisive.

The Need to Regionalize Research

The data compiled on the amount of human and financial resources devoted to research, in particular in Costa Rica and in Senegal, and on the three countries' dependence on foreign sources for funding research and related training clearly indicate that these countries cannot do everything themselves or work alone. The scientific machinery is becoming too expensive to run, even for some of the ICs (the smaller-

and medium-sized ones). This has convinced certain governments, in particular in Europe, to pool their resources in certain fields of research that require expensive equipment, e.g., nuclear physics and molecular biology.

The question of regional and international cooperation is especially acute since resources are limited and the economic/financial crisis is forcing certain countries to adopt policies of austerity and uninvolvement. Cooperation is of special importance in a continent such as Africa where 31 out of 50 countries have fewer than 5 million inhabitants, and only 2 (Egypt and Nigeria) have over 30 million. We might imagine Africa, or other continents, with minimal national structures and intense regional cooperation that would include division of labor, networks, and interstate research centers. We might be somewhat dubious about this type of reorganization as we think back to the mishaps of certain regional research institutes and the readily justified desire of countries to formulate their own research policies. Furthermore, this type of a strategy can only become genuinely productive if supported by well-consolidated national systems. We will come back to this question in our conclusions.

1. Referring to the "Interfuturs" scenario, Salomon and Lebeau (1988) suggest a typology with five Third Worlds and indicate that some countries would fit into more than one.

2. The fact that there is such a small number of scientists (66) working in the field of health, i.e., 4.5%, may be explained partly by the limited scope of the survey conducted in 1981. Actually, the Faculty of Medicine and Pharmacy had 170 teachers in 1984. It is difficult to estimate the number of teachers who also work on research.

3. Many ORSTOM scientists also work for the Centre de Recherche Océanographique at Dakar-Thiaroye (CRODT), which is a branch of ISRA. They provide scientific management and supervision and train young Senegalese for research work. This Franco-Senegalese cooperation is often cited as an example of success (Ruellan 1988).

4. Certain observers felt that the rapid Senegalization of the ISRA scientific staff meant converting many Senegalese technicians into research scientists. I was not able to check this.

5. By way of comparison, in chapter 2 we saw that in the United States women accounted for only 13% of the national scientific community; and in the field of agriculture, only 4% of the research scientists were women.

6. The cost of living for scholarship students abroad depends mainly on the country. The annual cost, in U.S. dollars, ranges from $3,000 in the USSR to $7,400 in the United States to $10,800 in Japan (Ramirez 1987). These figures do

not include registration and tuition fees nor travel expenses. With $4,500 as the average tuition fee for a semester in the United States and annual living expenses of $7,440 (actualized to cover inflation between 1985 and 1988), a master's degree taken in the United States, in other words, two and one-half years of study, would cost about $44,000; a doctorate requiring 4 years of study would "cost" $70,000. Just by way of comparison, a "maestria" at UCR in Costa Rica would only cost 227,600 colons (and this includes living expenses), which, at the 1987 exchange rate, comes to slightly less than $4,000, in other words, ten times less than in the United States (Ramirez 1987).

Chapter Seven

Conclusion

The results of the questionnaire sent to scientists in DCs, the analysis of their scientific output, and the comparative study of three national scientific communities confirm dependence of both the DC communities and their scientists on countries in the center and on international scientific communities. Outside sources are also heavily relied upon for training in research science, institution building, research financing, and other such fields. To a large extent, DC scientists use international scientific literature as their reference, choose research topics on the basis of essentially the same criteria as their colleagues in the center, and tend to select the same equipment that they grew accustomed to using during their Ph.D. studies in laboratories in the ICs.

But importing equipment manufactured in the "north" into the DCs of the "south," even with clear instruction manuals, is not enough to ensure equal quality service (Gaillard and Ouattar 1988). Similarly, scientists who studied in the "north" often discover that the subject of their thesis, their course curricula, knowledge, and experience are not directly applicable upon return to their home country. We have seen the blanket transfer to the "south" of models from the "north." This may complete the first steps of institution building, but institutionalization should not be synonymous with internationalization. It is becoming increasingly clear that applying major international criteria to scientific communities of the periphery, especially in the DCs, will not guarantee the latters' integration into international scientific communities. It may, quite to the contrary, detract from the relevance of research to local needs and problems.

The scientists of the national communities find themselves at the heart of a dilemma between their decision to participate in solving local problems and their attraction to models and reference systems more or less imposed by the international scientific communities. Some writers feel that the DC scientific communities are going through a transitional phase between initial establishment and integration into the international scientific community (Roche and Freites 1982). But for DC scientists, the "open sesame" into the international scientific community means adopting scientific standards and practices of the "north." This may mean, for instance, using English as the language of scientific communication and eliminating the use of national languages, choosing research themes that may interest international scientific journals, and thus neglecting national realities and potential national users of research results. What are the choices? We firmly believe that between the two extremes made up of the national versus the international course there are feasible midway strategies.

In our conclusion we will go further than reiterating the many results and the questions arising from them by looking at the political implications of these results, both as concerns the research assistance policies for the DCs and the national policies implemented by the DCs themselves.

The Uphill Emergence of a Scientific Attitude and Profession

The strategies adopted by the scientists are the result of negotiations carried out in a socioeconomic, cultural, and political environment that is not always very conducive to a scientific outlook and societal recognition of research science as a profession. This is in part due to the economic and financial situation that exists in most of the DCs. Moreover, the results and answers science has to offer are often considered unsatisfactory, because the DCs, hounded by growing marginalization in the world economic system, anguish for development. There are other, deeper reasons related to the endogenous cultures of the DC societies that do not have their own scientists to serve as examples.

In most traditional societies, learning is based mainly on routine and respect for knowledge held by the elders. Chapter 5 explains that critical views are not part of the tradition and cites the situation in Asia where tradition goes against the introduction of genuine scientific research and hinders normal scientific work. Respect, often blind respect for the teachings of old, makes it difficult to acquire a capacity for creative thought and learning through systematic observation. On the other hand, scientists trained abroad, in another cultural system, may easily be alienated from their original society. Another most legitimate question focuses on the effects the interaction between science and technology can have on the culture. Does the traditional society risk losing its cultural identity? This is the thesis of a Thai biochemist who felt that in certain cases preserving culture should be given higher priority than developing science and technology. He said, "Who can tell whether mankind is served better by building a power station where an old temple now stands?" (Yuthavong 1979).

As part of our case study, we tried to bring out the special importance of finding scientists who can be seen as examples during this era of newly emerging scientific communities. Three illustrious people, Clodomiro Picado in Costa Rica, Cheikh Anta Diop in Senegal, and King Mongkut, considered as the father of science in Thailand, are venerated by their scientific communities. Universities, research centers, and scientific awards have been named for them. Except for Clodomiro Picado (who held a state doctorate from the Sorbonne), we might ask what sort of scientists and scholars they actually were. Perhaps they were more like spiritual, or even religious and intellectual, leaders. This is a question worth exploring.

These scientific communities in quest of examples navigate in a scientific field that should not be mixed up with the intellectual field, the word *field* being taken in the sense of a gradient of forces within a specific environment. In 1987 in Ivory Coast, E. Fassin (personal communication) described the emergence of an intelligentsia, which, until recently, did not exist despite the considerable number of professionals working in education, journalism, and other white-collar jobs. The formation of this new group of "identity shapers," the intellectuals, was sparked by a combination of factors: a sudden economic hardship (with degree holders joining the ranks of the unemployed), new political classes, and a crisis in the model known as the Ivorian strategy. Certain scientists and university scholars, e.g., the economists, joined the group wearing the "halo" conferred by scientific work but often for reasons unrelated to their specific disciplines. This situation does not encourage the constitution of a scientific field separate from the intelligentsia. Up to a certain point, the early scientific controversies in the DCs and the appearance of the first leading scientists, especially in Africa, e.g., Cheikh Anta Diop in Senegal, can be traced back to an intellectual field with characteristics, such as autonomy from politics and science as a reference, that support the ambiguity.

Identifying a specific scientific "field" meant that society recognized a specialized group, a category of "scholars" that stood out socially from the population in general and also from the intellectuals, the elites, and the executive cadres. This meant splitting away from the practitioners, after taking distance from the state. Tripier (1984) has illustrated how the emergence of a profession occurs at the cost of the state and the market—at the cost of the market because it draws away members from a professional category and reclassifies them according to a recognized need for a new function, and the cost of the state because a domain that was previously controlled and run by the state becomes autonomous in defining its orientations and selling its results on the public market (except for the part of the market that is a state monopoly), thus reassigning former socioeconomic rules and regulations.

Up to the present, science in the DCs, especially in Africa, has been essentially controlled by government. The first step for the newly independent countries was to build up the state and its institutions. Education was given top priority in order to train civil servants for the state. This concept alone, without considering diploma qualifications, has often been used in assigning jobs; and careers have been constructed on successes or failures in the power struggle rather than through professional specialization. Because of this situation, it has been difficult for research science as a profession, or even as a vocation, to emerge in many DCs. We have seen that as a career it is not very appealing and is in urgent need of statutes. But isn't the problem being viewed backward? The absence of statutes is an indication that the profession has not really distinguished itself in society and that professional standards and representations do not gel properly.

Care must, however, be taken to avoid overgeneralizing on the basis of the three countries cited as examples in this paper. We have noted considerable differences between the three countries that are indicative of differences between their continents. In Africa, for instance, the social origin of the scientist is more indistinguishable than in the other continents. Conversely, in many countries in Asia, especially in Thailand and in India, the scientist's history traces back to the conversion to science of groups traditionally characterized by eminent relations to wisdom and/or power. In Thailand, for example, the royal family played, and is still playing, a very important role in the birth, growth, and dissemination of science. Princess Maha Chakri Sirindhorn, for instance, who recently received her Ph.D. in organic chemistry from the Mahidol University, is considered a symbol and often participates in public scientific events. With reference to the introduction of Western science in Bengal in the nineteenth century, K. Raj (1986) analyzes the historical vicissitudes that inspired members of the dominant Hindu cast, especially the Brahmans, to appropriate Western knowledge in India and form—and maintain to the present— the nucleus of a tightknit, socially esteemed scientific community.

Similarly, we were able to show that in Thailand and in Costa Rica the scientific spirit is developed more easily than in Senegal. We think encouragement should be given to activities such as national science days, science awards, science weeks for young people, annual conferences of national science associations, and also exhibits, science museums, and clubs that attract young people to science and scientific careers. Further, research scientists should be provided with the resources needed for their work. The recent example of two Jamaican scientists who received a national award for developing a treatment against glaucoma based on cannabis is very telling. After receiving the award, one was immediately hired by an American firm, the other by a German firm. Jamaica conferred the award but did not offer sufficiently attractive working conditions. In chapter 4 we saw that there was a correlation between job offers abroad and the number of years spent studying abroad. The place of education is almost decisive in shaping attitudes and scientific minds and in choosing research subjects and methods.

Education: At Home or Abroad?

Our survey of scientists and our case studies confirm that research training is heavily dependent on sources abroad. Notwithstanding increasing opportunities in many DCs, most students still go abroad, mainly to Europe and the United States, for their doctorate and often for their master's and sometimes even for their bachelor's degree. The only DC countries that have many doctoral candidates and enroll students from other DCs are India, Brazil, Mexico, Nigeria, Egypt, and the Philippines.

This dependency is incompatible with the creation of an independent scientific tradition and the emergence of a truly autonomous scientific community. It is becoming increasingly urgent to shift the "center of gravity" of doctoral-level education from the countries of the "north" to the "south." This will require cooperation between the northern host countries (which often offer scholarships) and the DCs themselves. The process will entail redefining aid policies (and the risk for the "north" countries to lose some of their influence) and, in many cases, DC education policies. Substantial sums have been spent by the countries that offered most of the scholarships, e.g., United States, France, United Kingdom, Canada, Australia. Thousands of potential research scientists were trained without any concern for their future roles in the national economic systems and more specifically in the national research systems. Corresponding sums could be used in DC universities to establish or strengthen doctoral courses in disciplines of national priority.

This change cannot take place overnight. During a transitional period, however long it may be, some compromise solutions will have to be applied. The "sandwich" formula that alternates fieldwork and data collection in the student's home country with courses and thesis preparation in the country of study seems most appropriate. Whenever possible, the student should defend his thesis in his home country. Original, innovative training policies adopted by certain schools of higher learning in DCs are worth looking at more closely and could serve as an example or a source of inspiration for other countries. One example is the Institut Agronomique et Vétérinaire Hassan II, in Rabat, Morocco, which applies the "sandwich" formula as part of its general education scheme. Innovations, such as the training-employment contracts in France, should also be used to guarantee newly trained scientists employment or integration into their national scientific communities. This type of a contract could provide a more satisfactory link between newly acquired scientific/technical skills and the profitable professional performance. Regional cooperation is also part of the answer and should be used to make up for shortcomings in the top-level national education systems. This is particularly true for various scientific and technical fields in many of the small DCs where training opportunities are still very limited.

Strengthening national academia would contribute to improving the structuring of the scientific communities as a result of added input from both the national scientific potential and the student body. This is essential if the actors on the scientific stage, from national leaders to Ph.D. candidates to regular students, are to keep up with science-in-the-making and remain up-to-date on progress in their disciplines. Student participation in research is doubly important. On the one hand, it teaches students that science is a method that can be used to state and solve problems. On the other, it contributes to making up for the shortage of research scientists, especially in field surveys, and providing opportunities to cover a large part, if not all, of the national territory.

All these proposals and recommendations do not exclude the possibility of pursuing highly specialized postdoctoral studies in an IC. But this usually requires outside assistance, especially financial backing, from countries of the "north."

Official Independence, Financial Dependence

All the DCs rely to some extent or another on research funding from abroad. Our case studies showed that the figures could vary from slightly under one-third (Thailand) to close to two-thirds (Senegal) of the research budget. Whatever the degree of dependency, the foreign funds are often vital to defray the operating costs of research. The share of foreign funding is especially critical in agriculture and health sciences. In agricultural research, Oram (1985) estimated that foreign aid paid for close to 40% of the R&D costs in the DCs. In certain countries, like Mali, Mozambique, Senegal, Lesotho, Swaziland, and Zambia, the figure is 70% or more. It is not always easy to distinguish the exact share of national and foreign funding. The fact that many universities do not have a research budget and that there is a large number of donors involved does not facilitate matters. Burkina Faso, for instance, which is a small African country, each year receives at least 340 foreign missions sent by governmental, multilateral, and international agencies to look at agricultural research problems (ISNAR 1983). The uncontrolled growth of foreign aid, especially in the 1970s, and the lack of coordination both among the aid dispensers and at the national levels created all sorts of problems, especially connected to the countries' capacity to absorb the aid.

Donor organizations providing research aid in DCs only started working on coordinating their efforts in the mid-1970s, probably as a reaction to the economic crisis. The follow-through was that in 1982 a group of donors (BOSTID, United States; GATE [German Appropriate Technology Exchange], Federal Republic of Germany; IFS; NUFFIC (Netherlands Universities Foundation for International Cooperation), Netherlands; and SAREC, Sweden), upon the initiative of IDRC of Canada, created IDRIS, the Interagency Development Research Information System, which serves as a data base for the group members by compiling descriptions of their respective DC-related research activities. In the start-up phase, IDRIS wanted to bring together a small number of organizations. However, it is open to new members, and its system is available to all outside users, especially the DCs. Similar initiatives have been taken by the World Bank through the CGIAR (Consultative Group for International Agricultural Research), which created a type of donors' club called SPAAR (Special Program for African Agricultural Research).

Of the many advantages to having donors work together, we need only mention greater transparency, better understanding of priorities and criteria for eligibility, and improved functioning. Working together facilitates harmonization and cooperation. The rules and standards for aid allocation by the donors are often ill known because they are not always explicit, tend to be confidential, or are simply not well publicized. Lack of cooperation is detrimental for the donors and even more so for the DCs that would greatly benefit from meeting together to share their attitudes to and uses of the aid they are offered, to assess advantages and drawbacks, and to discuss mechanisms that could be used nationally to obtain the most favorable aid conditions. In that context, the DCs would do well to consult the study by UCR's foreign financing section that has established a data base on research aid funding institutions (UCR 1985b). The most original part of this study is a confidential chapter that analyzes the difficulties with the various types of aid, i.e., delays, failures, lack of cooperation, problems of transferring funds and purchasing equipment, etc. The study also contains a classification of the donor agencies, based on their usefulness and efficiency as seen by the "recipient," and a data bank on the various organizations that provide research assistance. Paradoxically, many DCs fear that if there is too much coordination in research financing, the total amount of aid will be decreased. This is the attitude adopted by national agricultural research leaders in India, according to a recent external evaluation, and is shared by many science policymakers (Busch 1988).

We do not share this last point of view. On the contrary, we believe that more coordination in research financing should help control foreign aid and minimize the risk of national scientists undertaking research on themes that the home country does not prioritize. When the percentage of foreign financing is very high, the consequences suffered when funding agencies withdraw can be most serious. Many DCs know this situation well. Foreign aid will have to continue providing a goodly part of research funding, but the DCs would do well to reduce their relative importance by appealing elsewhere, e.g., the private sector. There are several mechanisms that could be used, e.g., a new company excise tax, a research incentives fund. Private—national or foreign—foundations could be persuaded to play a bigger role. This would be a further way of diversifying fields of operation and recipient institutions, since foreign public aid seems to be limited to the traditional fields such as agriculture and health, mainly treated within public research institutes, and thus does not encourage the development of new fields.

The Context of Institutions and the Role of Universities

The largest pool of research scientists works in the universities. But in most DCs, universities have little money for research, as we have seen, and are exceptionally reliant on foreign funding. A study conducted recently in Brazil concluded that close to 40% of the university research services could not survive without foreign aid (Schwartzman 1986). The universities offer the most qualified scientists, or at least the ones with the most diplomas, and produce the most mainstream science; the exact sciences are studied there and nowhere else. But the 1970s brought about deep changes in the universities, as we have seen. The quantum rise in enrollment, the increase in faculties and their gradual establishment outside the capital cities led to a type of Balkanization of the universities. Teaching conditions also changed. Often overloaded with elementary courses to teach, deprived of normal experimental facilities, and forced to take on outside jobs to make ends meet, the university research scientists find pitifully little time for research. The situation is critical. Is university research condemned to die out?

Many DCs were faced with and forced to skirt the problem. This was the case in India, for instance, which had a rigid university system under pressure from all sorts of minority groups and was marred by an ever-poorer level of recruitment and instruction. The reaction was to create a series of national university centers and technology institutes as "poles of excellence" based on other systems of recruitment and better links between teaching and research. Latin America favors another alternative solution by giving a leading role to private research centers for social science research. The expression of democracy that led to the creation of universities for the masses is not always compatible with the ambition to create productive research centers within these universities. This probably explains why more public research institutes were created or expanded.

These institutes, under tight government control, conduct research on priority, usually technical fields, mainly agriculture and health. More and more increasingly qualified research scientists work in these institutes, although their overall academic level is lower than that of their colleagues in the universities. They have fewer chances for refresher training and scientific communication. Because of their numerous assignments, little time is left for systematically recording research findings and even less for analyzing and preparing them for publication. Since these scientists are administratively assimilated to the civil servants and are in general more poorly paid than their university colleagues, they have to round out their income through outside jobs. This is a system that is more important for the institute scientists than for the university scientists because of their poor salaries and limited career opportunities. Here again, it is difficult to generalize. In some Latin American countries, like Brazil, we found the opposite to be true, at least in agricultural research. Brazilian scientists want to leave the university to work at EMBRAPA (National Agricultural Research Institute) where they have better wages and working conditions. And for reasons explained above, university scientists are authorized to top their income through part-time jobs outside the university.

The situation is worst in Africa where the turnover is especially high, the result being that much work is done sloppily, left incomplete, and not recorded. In Nigerian institutes, 20% to 50% of the scientific posts are vacant. The proportion is the highest among the senior scientists. The turnover rate is between 60% and 80% per five years, and the trend is growing worse (Idachaba 1980). ISNAR (1988, p. 55) has made a Pan-African study and concludes that "it still has not been possible to maintain a cadre of experienced agricultural researchers. The weakness of these experience profiles has had a profoundly negative impact on the institutional development and general productivity of NARS (National Agricultural Research System) in West Africa. Most significantly it has not been possible to develop experience and stable 'critical masses' of researchers in the key disciplines and commodities. Equally important, graduate recruits have frequently not received the kind of expert supervision and mentorship which is normally expected from experienced colleagues in scientific communities."

National systems are usually backed by private, often industrial reseach that, as we have seen, is practically nonexistent in Costa Rica (except for social science research), Senegal, and Thailand. There are only a few countries in Southeast Asia that allocate resources equal to those earmarked for R&D in the OECD countries. In most DCs the private sector goes no further than importing technology, partly to meet their immediate need for solutions and partly to compensate for their inadequate industrial R&D capacity. This does not mean that the private sector should conduct all industrial research. A certain capability exists within the universities, but there is not enough contact with the private sector to make use of it. Furthermore, relations between research institutes and universities are barely more developed than relations between universities and the private sector.

There are two major explanations for this dysfunctioning: first, the effects of doctoral studies abroad; and second, the inability of the universities to play the central role they should play to stimulate and structure the national scientific communities. Doctoral studies, if pursued in the DCs, would be designed to react more to the economic realities and needs of both the private sector and the public research centers. Contacts students establish through the university during their doctoral studies could more easily be maintained after they have graduated and gone to work for public research centers or the private sector. More incentives should be adopted to promote a better integration of research activities within the different research institutions (universities, private and public centers, etc.).

Incentive measures might include encouraging researchers from various institutes to work together on joint projects, encouraging exchanges between staff members from various types of institutes for varying periods of time, authorizing institute scientists to attend refresher training courses at the university, urging the national journals (which are often published by the universities) to adopt a more open editorial policy, organizing research seminars regularly and inviting researchers from various types of institutes, making national journals more visible, and making unpublished documents more readily available by creating a central index system for reference materials.

Problems of Communication

The last suggestion introduces the problem of scientific communications. Answers to the questionnaire indicated that communications between DC research scientists were infrequent and were often restricted to contacts with colleagues in the same institute or even, often, in the same department or research unit. Contacts with scientists working elsewhere within the country are as infrequent as with scientists working in countries of the "north." Scientists who spent the longest time studying abroad had the most contact with foreign scientists. Similarly, they tended to depend more on work published in the ICs and made little use of what their colleagues at home or in other DCs published.

DC scientists often cite colleagues in ICs but not vice versa, even when works by DC scientists are published in well-read international journals. This behavior seems to be the result of a rather widespread, although difficult to prove, conviction among DC scientists that quoting works published by colleagues from ICs brings more credit to their own work. However, scientists from the ICs seem to feel that the work published by their DC colleagues is insignificant. This attitude can have unexpected consequences, as is shown in the example published by two colleagues from ORSTOM (Chatelin and Arvanitis 1988a) concerning two young Philippino researchers who published their findings on the sensitivity of maize hybrids to a fungus called *Helmintosporium maydis*. Four years later, an American scientist working on the same question discounted the warning and claimed that in the Philippines hybrids were not resistant enough and local agricultural practices were bad. That was in 1965. In 1970 the United States suffered tremendous maize crop destruction due to *Helmintosporium*. This example should make researchers in the ICs more circumspect, wary of presumptive judgments and undue simplifications that tend to consider DC science as "bad" and international science, dominated by the countries of the "north," as "good."

Fortunately, blanket judgments are not systematic and are not the only explanation for the widespread misunderstanding of research work conducted in the DCs. The lack of visibility and accessibility of DC science aggravates the problem of unsatisfactory dissemination. We mentioned that DC scientists often publish their findings in national journals. The data Léa Velho (1985) compiled in Brazil showed that Brazilian scientists not only chose national journals but went further and often chose in-house publications. This happens in many DCs. Circulation becomes even more of a problem when the language of publication is a local tongue, unknown beyond national borders, e.g., Thai. These publications are very seldom listed in international data bases. Furthermore, the number of copies printed is very limited, and the radius of circulation is small. They are not published regularly and often, because of financial difficulties, stop appearing after the first few issues. What can be done? Again, there are solutions, but they require resources.

In Senegal, as we have already mentioned, ISRA recently published an analytical bulletin of its scientists' work. This document, and the data base it inspired, provide a new degree of visibility and are being used as tools to structure scientific communications at the national level and, hopefully, at a later date, will be expanded to the regional and international levels. This could be an example for other national institutions and serve in a nationwide common reference index system. After the base has been established, it could be consulted via the telecommunications network. Procuring a terminal that can access data bases should be a priority for unequipped scientific institutions. Some data bases such as FAO's AGRIS (international information system for the agricultural sciences and technology) are intended for the DCs and contain large quantities of nonconventional literature such as university theses, reports, and unpublished papers.

Modern microcomputer technology also has an important role to play. Far be it from us to contribute to the mystifying utopia that pretends that the information revolution will provide a formidable shortcut to solving all the ills of underdevelopment. But we do need to recognize the often accelerated availability of new tools in many DC research institutes. This was observed by the editors of a journal entitled *Tropical Animal Production* that is computer printed and now called *Tropical Animal Production for Rural Development*. The Instituto Mayor Campesino in Buga, Columbia, in 1988, wants it to become a research journal to promote communications among scientists and decision makers concerned with rural development in the Third World. The journal is produced and distributed in diskette form and printed locally. DC institutions and scientists will have access to the journal free of charge. To receive the journal merely requires sending an unused diskette to the distribution centers in Columbia or England. Here again, the success of the project partly depends on assistance from the "north" in providing microcomputers to unequipped DC institutes.

National institutions in charge of science policies and donor organizations should also continue furthering the establishment of both formal and informal networks by convening national and regional conferences and by publishing and widely distributing conference proceedings. Scientists in the ICs should also be given opportunities to participate in meetings organized by the DCs. Considering how poorly scientific information is circulated in the DCs, seminars and conferences are still probably the most effective tool for establishing durable scientific contacts. National leaders and foreign aid organizations should realize this and provide resources for DC scientists to honor invitations to conferences abroad. All too often they cannot fight through the obstacle course lined with problems of exit visas, lack of foreign currency, depleted travel budgets, and vague suspicions by the administration that the trip is a joy ride or an exaggerated privilege for a senior civil servant.

Relations between research scientists and scientific communities in other DCs, especially at the regional level, need to be revised. Especially for the small countries, training, funding programs, and organizing conferences are too demanding to shoulder at the national level.

Critical Mass and Regional Cooperation

Thus, many DCs, because of their small size and/or lack of resources, cannot enjoy scientific and technical autonomy and are tempted to favor the solution of regional cooperation, either through networks or regional research centers. Proposals along these lines have been discussed at great length in regional and international conferences on science and technology. This was the case at the First Conference of Men of Science in Africa (Brazzaville, June 1987) and at the second CASTAFRICA (Conference of Ministers Responsible for the Application of Science and Technology to Development in Africa) conference, which, in Tanzania, in July 1987, brought together ministers of science and technology in Africa. We have to recognize, unfortunately, that participating states are often willing to allocate funds to create networks and regional scientific organizations but then do not allocate funds to run them.

Furthermore, there are certain problems, especially related to health, that have to be solved at the regional level because of their scope and nature, e.g., disease-bearing insects, while others, especially in agriculture, are more site-specific and have to be researched as such. Results deemed satisfactory for one area may not be for another. This is why the network system, in a more or less institutional form, seems to offer a less stringent answer that is better adapted to the requirements of regional, continental, and international cooperation. This type of cooperation is not new in the field of research. What is new is the recent proliferation of networks, especially in sub-Saharan Africa, and the attendant need for regional cooperation. SPAAR has inventoried over 60 research networks in this region of the world devoted to subjects such as information exchange, scientific consultation, and joint research.

There are new systems and fashions in multilateral partnerships that, in time, may cause national systems to split into two subsystems, one better endowed because of its support from the networks and the other left to fare for itself. The underlying risk will be that network-supported programs all take root in a small number of scientifically more advanced countries. Actually, regional scientific cooperation can only exist if the participating countries have adequate national scientific policies and capabilities. Otherwise, national scientific choices may be sacrificed on the altar of regional or international cooperation. In other words, we cannot ignore the need to strengthen national research systems.

Redefining National and Donor Policies

We have seen that in many cases ample, although not always adequate, human and financial resources are made available to create institutions. The key problem is one of quality. Less attention is needed for size and resources than for good functioning and effectiveness. Actions need to be sited on the scientific field, e.g., detailed identification of themes with a future, assistance in training teams, continued support for the more promising ones, assistance in communications through written media such as national journals, and oral media such as conferences. This will help consolidate the national scientific fabric and give it an international dimension.

On the DC side, the vital component is the national scientific policy, whose effectiveness will depend on an improved professional status for the research scientists and increased autonomy for the institutions. This seems to be the most appropriate place to stress the need for reliable, regularly updated statistics. Tools of analysis such as indicators or inventories of scientific and technical potential, a scientific plan or budget, or even a definition of priorities all need an adequate statistical base. There is no use speaking about a science policy or research management unless adequate support is provided.

For the aid policymaker, this means maintaining a general overview but relying less on aid packages and more on scientific operators. The type of action involved exceeds the capacity of the global administrative operators who are used to handling huge budgets. Innovations for institutional structures and new work parameters will be needed. The most vital steps will be helping national scientific communities to emerge, developing internal standards and external capacities for project formulation, negotiation, and orientation. The awards system will have to be better adapted to the needs of development and the requirements of research in cooperation.

Developing Science Takes Time

Some 25 years ago, most people thought that the technologies needed to develop the Third World were available. Some people even thought that the DCs were starting the march to development at a favorable moment in history and that all they needed to do to catch up was to shop in the supermarket of technologies for ready-made solutions designed by the ICs. The general thought was that there was a direct, linear relation between basic research, applied research, technological development, and, downstream, economic growth. Knowledge was expected to flow through the system like water in a pipe. Quixotically, it was thought that injecting proper investments and trained scientists at the entrance of an R&D system guaranteed economic development at the exit. True, science and technology brought World War II to an end in six years and put a man on the moon in eight; but the facts show that they have not been able to win the battle against the poverty rampant in most DCs, even though the DC populations have been able to benefit from tangible progress, especially in certain fields such as health and nutrition.

We now know that the relations between science, technology, and development are much more complicated and that there is no shortcut to solving development-related problems. Developing science is a lengthy enterprise. Even under conditions more favorable than those facing most DCs in the world of today, countries like the United States and Japan needed 50 years to be able to compete with European countries. A study on the process of industrializing scientific research in Western countries highlights the decisive factor as being time, the time needed for science to find its legal niche in the social system.

Most DCs are still in the institutionalization and professionalization stage. The initial institutional structure has been created, but its material expression, scientific research, has not yet been institutionalized, i.e., been recognized as a full-fledged component of society. Unlike the Western countries, in the DCs the various steps in scientific development overlap each other.

The questionnaire and interviews have shown that scientists who have studied abroad find it very difficult to fit into their national scientific communities and must accept numerous sacrifices. There are certain prerequisites to making scientific investments "profitable." It is not enough just to lay down an institutional structure, train good scientists, and give them proper supplies. The scientists need to be able to find their place in a tightknit, lively scientific community that has its own legitimate seat in society.

Questionnaire for IFS Grantees

1. This questionnaire is intended for all IFS grantees, even those no longer receiving IFS support for their research work.

2. Try to answer all the questions.

3. Definitions of words marked with an asterisk are given in the appendix to this questionnaire.

4. The squares to the right are not to be filled in.

Do not complete items 1 through 8.
1. Answer to the questionnaire (yes/no). □
2. Agreement number (001 to 844). □□□
3. Country in which the grantee is working. □□□
4. Year of first grant (between '74 and '84). □□
5. Research area (1 to 7). □
6. Number of grants (1 to 4). □
7. Total amount received from IFS by 31 Dec. □□□
   1984 in thousands of Swedish SEKs.
8. IFS support active or completed. □

I—Civil Status and Education

9. Family name _____

10. First and middle names _____

11. Names and addresses of your home institution _____

_____

_____

12. Citizenship _____

13. Sex: Male □   Female □ □

14. Year of birth: 19 __ □□

15. What was your father's principal occupation when you were □
    16 years old? (If deceased or retired when you were 16, indicate
    previous occupation.)

_____

16. Where did you live when you were 16 years old? □

☐ in your national capital?
☐ in a town other than the capital with over 50,000 inhabitants?
☐ in a town of between 2,500 and 50,000 inhabitants?
☐ in a town of under 2,500 inhabitants?
☐ in a village or rural area?

17. Are you married? Yes ☐   No ☐   ☐

18. How many children do you have?_____   ☐

19. If you are married, what is your wife's principal occupation? _____ ☐☐

_____

20. Higher education.
    In chronological order list the diplomas you obtained, starting with the
    B.S. level. Indicate the scientific area of specialization and the name and
    country of the educational establishments you attended.

    A. Diploma (or degree) _____   ☐
       Field of studies _____   ☐☐
       Year _____   ☐☐
       Establishment _____   ☐
       Country _____   ☐☐☐

    B. Diploma (or degree) _____   ☐
       Field of studies _____   ☐☐
       Year _____   ☐☐
       Establishment _____   ☐
       Country _____   ☐☐☐

    C. Diploma (or degree) _____   ☐
       Field of studies _____   ☐☐
       Year _____   ☐☐
       Establishment _____   ☐
       Country _____   ☐☐☐

    D. Diploma (or degree)_____   ☐
       Field of studies _____   ☐☐
       Year _____   ☐☐
       Establishment _____   ☐
       Country _____   ☐☐☐

    E. Diploma (or degree) _____   ☐
       Field of studies _____   ☐☐
       Year _____   ☐☐
       Establishment _____   ☐
       Country _____   ☐☐☐

21. If you wrote a Ph.D. thesis, give your supervisor's name and address
    and the title of your thesis:_____

_____

_____

_____

_____

22. Languages.
Mother tongue: _____   ☐
Please complete the following table:

	Read	Speak	Write
Languages	E NE	E NE	E NE
_____			
_____			
_____			
_____			
_____			

☐☐☐☐
☐☐☐☐
☐☐☐☐
☐☐☐☐
☐☐☐☐

E = Easily   NE = Not Easily

23. How many years did you spend outside of your country for
higher education and training, including postdoctoral
studies, etc.? ____ years   ☐ ☐

24. How many years did you spend abroad altogether? ____ years   ☐ ☐

25. Why did you choose to become a scientist or a research worker?
How important were the following considerations in your choice? Please
rate each criteria by circling one number from "essential" to "not
important at all."

> 1 = Essential
> 2 = Very important
> 3 = Moderately important
> 4 = Not very important
> 5 = Not important at all

- Social status................1 2 3 4 5   ☐
- Job security ....1 2 3 4 5   ☐
- Remuneration ....1 2 3 4 5   ☐
- Career prospects ....1 2 3 4 5   ☐
- Intellectual stimulation ....1 2 3 4 5   ☐
- Social utility ....1 2 3 4 5   ☐
- Parental influence ....1 2 3 4 5   ☐
- Influence of a professor ......1 2 3 4 5   ☐
- Other ....1 2 3 4 5   ☐

II—Position

26. Job.
1. Present position (title) _____   ☐
    Starting date (year) _____   ☐

Amount of time devoted to
Teaching ____% Administration ____%   □□ □□
Research ____% Extension ____%   □□ □□
Other (specify) ____%   □□

2. Title of your position at the time you applied for your first IFS grant:
_____   □

Amount of time devoted to
Teaching ____% Administration ____%   □□ □□
Research ____% Extension ____%   □□ □□
Other (specify) ____%   □□

3. Title of your first professional position: _____   □

27. What is at present your main field of science, e.g., agronomy,
chemistry, zoology, etc.? _____   □□

28. Since the beginning of your research career, have you substantially
changed your scientific orientation/research subjects?
    Yes □   No □   □

29. If "yes," what motivated this change? _____   □
_____
_____
_____

30. Do you consider that the salary you receive as a scientist is adequate to
support yourself and, if applicable, your family?
    Adequate □   Inadequate □   □

31. Compare your salary as a scientist or teacher-cum-scientist with the
minimum wage in your country by indicating how many
times more you get.
    ____ times more   □□

32. If you are obliged to have side jobs to support yourself and, if
applicable, your family, indicate how many hours they
require per week.
    ____ hours   □□
Specify the nature of your side jobs. _____
_____   □

III—Research Choice and Objectives, Working Relations

33. How would you say the activities in your department or research unit
are divided up? What do you think the time distribution should be?
Indicate percentages of time.

Present		Ideal	
____%	Teaching	____%	□□□□
____%	Basic research	____%	□□□□

_____%   Applied research   _____%   □ □ □ □
_____%   Development   _____%   □ □ □ □

34. How would you characterize the distribution of time in your IFS-supported program? What do you think it should be? Indicate percentages of time.

Present		Ideal					
_____%	Teaching	_____%	□	□	□	□
_____%	Basic research	_____%	□	□	□	□
_____%	Applied research	_____%	□	□	□	□
_____%	Development	_____%	□	□	□	□

35. When you prepared your first research grant application to IFS, how important were the following considerations in your choice of research problems? Please rate each criterion by circling a number from "essential" to "not important at all."

1 = Essential
2 = Very important
3 = Moderately important
4 = Not very important
5 = Not important at all

Criteria for choice:

1 2 3 4 5 Potential contribution to scientific theory   □
1 2 3 4 5 Likelihood of clear empirical results   □
1 2 3 4 5 Potential creation of new methods, useful materials, and
          devices   □
1 2 3 4 5 Potential marketability of the final product   □
1 2 3 4 5 Some funding available from your institution's budget   □
1 2 3 4 5 Some funding available from sources other than your
          institution   □
1 2 3 4 5 Length of time required to complete the research   □
1 2 3 4 5 Possibility to publish in professional journals   □
1 2 3 4 5 Possibility to prepare M.S. or Ph.D. thesis (research
          subject)   □
1 2 3 4 5 Priority research areas of the IFS program   □
1 2 3 4 5 Availability of research facilities   □
1 2 3 4 5 Currently a "hot" topic   □
1 2 3 4 5 Colleague's approval   □
1 2 3 4 5 Your reputation among other scientists doing similar
          research   □
1 2 3 4 5 Pleasure of doing this kind of research   □
1 2 3 4 5 Importance to society   □
1 2 3 4 5 Scientific curiosity   □
1 2 3 4 5 Customer demand   □
1 2 3 4 5 Feedback from extension personnel   □

1 2 3 4 5 Priorities of your research institution   ☐
1 2 3 4 5 Other_____   ☐

36. In choosing your research subject, in what ways did the following
    people influence you? Indicate the degree of influence by circling one
    number from "essential" to "not important at all."

    1 = Essential
    2 = Very important
    3 = Moderately important
    4 = Not very important
    5 = Not important at all

    1 2 3 4 5 Your immediate supervisor   ☐
    1 2 3 4 5 A colleague in your department   ☐
    1 2 3 4 5 A colleague in another department at your institution ☐
    1 2 3 4 5 A colleague at another institution   ☐
    1 2 3 4 5 A research assistant/technician   ☐
    1 2 3 4 5 A graduate student   ☐
    1 2 3 4 5 A professor other than your Ph.D. supervisor   ☐
    1 2 3 4 5 Your Ph.S./Ph.D. supervisor   ☐
    1 2 3 4 5 The director of your research facility   ☐
    1 2 3 4 5 A client or potential user   ☐
    1 2 3 4 5 An IFS scientific adviser   ☐
    1 2 3 4 5 Another IFS grantee   ☐
    1 2 3 4 5 A member of the IFS staff   ☐
    1 2 3 4 5 The research review committee   ☐
    1 2 3 4 5 Other (specify)

37. How often do you communicate with the following people regarding
    your research? (Please circle one for each line: 1 = never, 2 = rarely,
    3 = annually, 4 = monthly, 5 = biweekly, 6 = weekly, and 7 = daily.)

    1 2 3 4 5 6 7 Scientists in your department   ☐
    1 2 3 4 5 6 7 Scientists outside your department in your institution ☐
    1 2 3 4 5 6 7 Scientists from other institutions in your country   ☐
    1 2 3 4 5 6 7 Scientists outside your country   ☐
    1 2 3 4 5 6 7 IFS scientific advisers   ☐
    1 2 3 4 5 6 7 IFS grantees in your country   ☐
    1 2 3 4 5 6 7 IFS grantees outside your country   ☐
    1 2 3 4 5 6 7 Your Ph.S./Ph.D. supervisor   ☐
    1 2 3 4 5 6 7 IFS secretariat   ☐
    1 2 3 4 5 6 7 Other funding agencies   ☐
    1 2 3 4 5 6 7 Extension staff   ☐

38. To carry out your research project, do you work alone or with other
    scientists?
         Alone ☐   With other scientists ☐   ☐

39. Do you correspond by mail on scientific problems with foreign scientists?

Yes ☐   No ☐   ☐

IV—Research Funding

40. What was your annual research budget (excluding salaries) at the time you applied for an IFS grant, to the nearest U.S. $1,000?

U.S. $ _____   ☐ ☐

41. What was the annual research budget (excluding salaries) last year (or during the last year you received IFS support, if your grant has been terminated), to the nearest U.S. $1,000?

U.S. $ _____   ☐ ☐

42. What were your sources of research funds in percentages (excluding salaries) last year (or during the last year you received IFS support, if your grant has been terminated)?

Home institution. . . . . _____%   ☐ ☐
IFS . . . . _____%   ☐ ☐
National funding organization (name) . . . . _____%   ☐ ☐
International funding organization (name) . . . . _____%   ☐ ☐
Other (specify) . . . . _____%   ☐ ☐
Total   100%

43. Would you have pursued your research if IFS funding had not been made available?

(1) ☐ Yes, other support would have been available   ☐
(2) ☐ Yes, even without other support   ☐
(3) ☐ Yes, but on a reduced scale   ☐
(4) ☐ Yes, but in a substantially different form   ☐
(5) ☐ No   ☐
(6) ☐ Other (specify)   ☐

44. Since becoming an IFS grantee, has it become easier for you to obtain:

	Yes	No	
- additional funding from your institution	☐	☐	☐
- additional funding from a national funding institution	☐	☐	☐

If yes, give name _____

V—Scientific Literature, Publications, and Attendance to Conferences

45. After becoming an IFS grantee, how many of each of the following types of publications have you authored or coauthored?

Authored		Coauthored		
_____	Articles in scientific journals	_____	☐ ☐	☐ ☐
_____	Articles in conference proceedings	_____	☐ ☐	☐ ☐
_____	Books	_____	☐	☐
_____	Chapters in books	_____	☐	☐
_____	Bulletins	_____	☐	☐

____	Reports	____	☐ ☐
____	Other (specify)	____	☐ ☐

46. Indicate in what scientific journals you have had articles published
(as an author or coauthor).   ☐

a _____ f _____
b _____ g _____
c _____ h _____
d _____ i _____
e _____ j _____

47. To what journals do you subscribe or have easy and regular access?   ☐

a _____ f _____
b _____ g _____
c _____ h _____
d _____ i _____
e _____ j _____

48. Do you have easy and regular access to bibliographic catalogs such as
"Current Contents"?
    Yes ☐   No ☐
    If "Yes," which ones?_____

49. How important is the following literature in your research? Circle one
number from "essential" to "not important at all."
    1 = Essential
    2 = Very important
    3 = Moderately important
    4 = Not very important
    5 = Not important at all

    1 2 3 4 5 National journals in your field   ☐
    1 2 3 4 5 National journals in related fields   ☐
    1 2 3 4 5 Foreign journals in your field   ☐
    1 2 3 4 5 Foreign journals in related fields   ☐
    1 2 3 4 5 Books and monographs   ☐
    1 2 3 4 5 Research bulletins   ☐

50. Since becoming an IFS grantee, how many scientific conferences have
you attended?

Total		With financial support from IFS	
____	Within your country	____	☐ ☐ ☐
____	Outside your country	____	☐ ☐ ☐

51. Do you have access to bibliographic data banks?
        Yes ☐   No ☐   If yes, which one(s)?
    a _____ c _____
    b _____ d _____

52. During the last 12 months, about what percent of your research time did you spend:
 - in the office ___%   □□
 - in the library ___%   □□
 - in the field ___%   □□
 - in the laboratory ___%   □□
 - in the computing facility ___%   □□
 - in animal and plant experimental areas ___%   □□
 - elsewhere (specify) ___%   □□

VI—Supporting Activities: Technical Assistance, Purchase and Maintenance of Equipment, Evaluation

	The year of your first IFS grant	Last year		
53. Number of scientists you supervised:	_____	_____	□	□□
54. Number of technicians you supervised:	_____	_____	□	□□
55. Number of field-workers you supervised:	_____	_____	□	□□

56. After becoming an IFS grantee, did it become easier for you to obtain scientific and technical assistance from your institution?
 Yes □   No □   □

57. Do you or could you have easy access to a vehicle for your research work?
 Yes □   No □   □

58. Do you or your institution receive catalogs of research equipment and supplies regularly?
 Yes □   No □   □

59. What is the publication date of the more recent catalogs on research equipment and supplies currently available to you?
 19___   □

60. Do you have technicians in your institution who can install, maintain, and/or repair research equipment?
 Yes □   No □   □

61. The last time you had a piece of equipment that had to be repaired abroad or by traveling technician from abroad, how long did it take to get it repaired?
 ___ month(s)   □□

62. Do you make the equipment purchased with IFS grant funds available for

Yes No

- teaching activities   ☐ ☐   ☐
- other research projects in your research units or
  department   ☐ ☐   ☐
- other research projects in other departments of your
  institution   ☐ ☐   ☐
- other research projects in other institutions   ☐ ☐   ☐

63. Are research activities adequately encouraged in your country?
    Adequately ☐   Inadequately ☐   ☐

64. How do you perceive your government's attitude toward research?
    Indicate the attitude by circling one number between "negative" (1)
    and "very positive" (7).
    negative   1 2 3 4 5 6 7   very positive   ☐

65. Is your research work evaluated regularly?
    Yes ☐   No ☐   ☐

66. If yes, by whom?
    _____   ☐

67. Which criteria are used to grant promotions to scientists in your
    country?
    _____
    _____
    _____   ☐

68. Have you had a sabbatical leave:

                                Yes No
    - since the beginning of your career?   ☐ ☐   ☐
    - since you became an IFS grantee?   ☐ ☐   ☐
      If yes, which year? 19____   ☐ ☐
      in which country? _____   ☐ ☐ ☐
      for how long? ____ months   ☐ ☐

69. Have you been offered employment abroad?
                                  Yes No
    - since the beginning of your research career? ☐ ☐   ☐
    - since you became an IFS grantee?   ☐ ☐   ☐
      If yes, in which country?_____   ☐ ☐ ☐
    - Did you accept the offer?   ☐ ☐   ☐

VII—Implementation, Use of Research Results, Recognition, Awards

70. Excluding your own discipline, do you believe that your IFS-supported
    research and publication(s) have already or will directly or indirectly
    benefit any of the following? (Circle the corresponding numbers on each
    line from 1 = not at all to 5 = a great deal.)

Has or does benefit		Will benefit	
1 2 3 4 5	Other scientific disciplines Specify: _____	1 2 3 4 5	☐ ☐
1 2 3 4 5	Small farmers	1 2 3 4 5	☐ ☐
1 2 3 4 5	Large farmers	1 2 3 4 5	☐ ☐
1 2 3 4 5	Agri-business	1 2 3 4 5	☐ ☐
1 2 3 4 5	Rural area inhabitants	1 2 3 4 5	☐ ☐
1 2 3 4 5	City dwellers	1 2 3 4 5	☐ ☐
1 2 3 4 5	General public	1 2 3 4 5	☐ ☐
1 2 3 4 5	Local or state agencies	1 2 3 4 5	☐ ☐
1 2 3 4 5	Local industries	1 2 3 4 5	☐ ☐
1 2 3 4 5	Other countries in the region	1 2 3 4 5	☐ ☐
1 2 3 4 5	Foreign or international industries	1 2 3 4 5	☐ ☐
1 2 3 4 5	Other	1 2 3 4 5	☐ ☐

71. If your research results are being implemented or have already been ☐
implemented, briefly describe how and which national or international
agency/agencies helped.

_____

_____

_____

_____

_____

_____

72. How many people have you trained?

	Since the beginning of your career	Since you became an IFS grantee		
Technicians	_____	_____	☐ ☐	☐ ☐
M.S. students	_____	_____	☐ ☐	☐ ☐
Ph.D. students	_____	_____	☐ ☐	☐ ☐
Post doctoral students	_____	_____	☐ ☐	☐ ☐
Other (specify)	_____	_____	☐ ☐	☐ ☐

73. Have you received any prizes/awards in recognition of your outstanding
contributions in the field of your scientific specialization?

	Yes	No	
Before you became an IFS grantee	☐	☐	☐
Since becoming an IFS grantee	☐	☐	☐

74. What is, according to you, the main factor holding back your research
work? ☐

_____

_____

_____

_____

75. Are there any research interests or projects you would like to pursue that you have not yet been able to?

    Yes ☐   No ☐   ☐

76. If yes, specify and indicate constraints to their implementation.   ☐

    _____
    _____
    _____
    _____

77. What is your future career goal?

    ☐ Continued scientific career   ☐
    ☐ Politics   ☐
    ☐ Private business   ☐
    ☐ Development programs   ☐
    ☐ Other (specify)   ☐

No questionnaire of this type can adequately cover all the points considered relevant by individuals with so wide a range of interests, so please use the next page for any additional comments you may have.

Thank you for your cooperation. Please return the completed questionnaire together with a complete list of anything you have had published in the original language of publication (names of coauthors, titles of scientific journals, number of pages, dates, etc.) to IFS by *31 March 1985*.

_____

Additional remarks:

Countries in Which Respondents Are Working

Africa	Latin America	Asia & Middle East
Burkino Faso	Argentina	Bangladesh
Burundi	Barbados	China
Cameroon	Brazil	Hong Kong
Congo	Chile	India
Egypt	Colombia	Indonesia
Ethiopia	Costa Rica	Iran
Gabon	Cuba	Jordan
Ivory Coast	Dominican Rep.	Korea (Rep. of)
Kenya	Ecuador	Laos
Madagascar	Guatamala	Malaysia
Malawi	Haiti	Pakistan
Mali	Jamaica	Philippines
Mauritius	Mexico	Singapore
Morocco	Nicaragua	Sri Lanka
Niger	Panama	Thailand
Nigeria	Peru	Taiwan
Senegal	Uruguay	Vietnam
Seychelles	Venezuela	
Sierra Leone		
Somalia		**Pacific**
Sudan		
Tanzania		Fiji
Togo		Papua New Guinea
Tunisia		Solomon Islands
Uganda		Vanuatu
Zaire		
Zambia		
Zimbabwe		

Tables

Table A1. Marital Status by Age

Age in 1986	Married		Unmarried		Total	
50 yrs or over	36	92.30%	3	7.70%	39	100%
45 to 49 yrs	74	96.10%	3	3.90%	77	100%
40 to 44 yrs	119	90.84%	12	9.16%	131	100%
35 to 39 yrs	123	81.45%	28	18.55%	151	100%
30 to 34 yrs	52	67.53%	25	32.47%	77	100%
29 yrs and under	3	30.00%	7	70.00%	10	100%
Total	407	83.92%	78	16.08%	485	100%

Table A2. Number of Children by Age

Age in 1986	0	1	2	3	4	5	6	7	8	Total
50 yrs or over	3	3	9	11	8	3	1	0	1	39
			23.07	28.20	20.51					8.04%
45–49 yrs	6	8	24	21	8	5	5	0	0	77
			31.16	27.27	10.38					15.88%
40–44 yrs	17	14	49	28	16	6	1	0	0	131
			37.40	21.37	12.21	4.6				27.01%
35–39 yrs	40	26	39	23	17	4	0	2	0	151
	26.49	17.21	25.82	15.23	11.25					31.13%
30–34 yrs	36	18	18	4	1	0	0	0	0	77
	46.75	23.37	23.37							15.88%
29 yrs or under	7	2	1	0	0	0	0	0	0	10
	70.00	20.00	10.00							2.06%
Total	109	71	140	87	50	18	7	2	1	485
	22.47	14.65	28.87	17.94	10.30	3.71	1.44	0.41	0.21	100%

Table A3. Number of Children by Years Spent Abroad

Years spent abroad	0–3 children		4–8 children		Total	
0	65	16.09%	8	10.53%	73	15.21%
1–2	94	23.27%	18	23.68%	112	23.33%
3–4	97	24.01%	20	26.32%	117	24.37%
5–9	128	31.68%	23	30.26%	151	31.46%
10–20	20	4.95%	7	9.21%	27	5.63%
Total	404	100%	76	100%	480	100%

Table A4. Number of Years Spent Abroad by Father's Profession

Father's Profession	0	1–2	3–4	5–9	10–20	Total
Agriculture	22	35	36	41	8	142
	15.49%	24.65%	25.35%	28.87%	5.63%	100%
Self-employed	16	28	29	30	6	109
	14.68%	25.69%	26.60%	27.52%	5.50%	100%
Craftsman or trade	10	19	21	33	4	87
	11.49%	21.84%	24.14%	37.93%	4.60%	100%
Executive	4	15	14	18	3	54
	7.40%	27.78%	25.92%	33.33%	5.55%	100%
Employee	8	7	9	11	1	36
	22.22%	19.44%	25.00%	30.55%	2.78%	100%
Laborer	2	5	4	6	0	17
	11.76%	29.41%	23.53%	35.29%	0.00%	100%
Service staff or watchmen	5	2	1	3	2	13
	38.46%	15.38%	7.69%	23.08%	15.38%	100%
Other categories	2	1	1	2	0	6
	33.33%	16.66%	16.66%	33.33%	0.00%	100%
Total	69	112	115	144	24	464
	14.87%	24.14%	24.78%	31.03%	5.17%	100%

Table A5. Choice of Research Subject, Importance Given to Criterion: "Potential Contribution to Scientific Theory" by Number of Years Spent Abroad

No. of Years Spent Abroad	Essential or Very Important		Moderately to Relatively Unimportant		Not Important at All		Total	
0	39	56.52%	25	36.23%	5	7.25%	69	100%
1–2	57	52.29%	47	43.12%	5	4.59%	109	100%
3–4	58	50.43%	49	42.61%	8	6.96%	115	100%
5–9	83	56.08%	53	35.81%	12	8.11%	148	100%
10–20	13	52.00%	11	44.00%	1	4.00%	25	100%
1–20	211	53.15%	160	40.30%	26	6.55%	397	100%
Total	250	53.68%	185	39.70%	31	6.65%	466	100%

Table A6. Relative Influence of Various People in Choice of Research Topic

Rank	Person	Average[a]	Rating
1	Your immediate superior		3.31
2	The director of your research unit		3.52
3	A customer or potential user		3.60
4	A colleague from another institute		3.67
5	A colleague in your department		3.69
6	Your thesis supervisor		3.72
7	Your IFS scientific adviser		3.78
8	A professor other than your thesis supervisor		3.81
9	A research evaluation committee		3.94
10	A colleague from another department in your institute		4.03
11	A research assistant		4.16
12	A member of the IFS Secretariat		4.26

[a]Average based on a 5-figure scale (1 = essential, 2 = very important, 3 = relatively important, 4 = relatively unimportant, 5 = not important at all).

Table A7. Number of Years Spent Abroad by Age

Age	0	1–2	3–4	5–9	10–20	Total
25–29	4	3	2	1	0	10
	5.48%	2.65%	1.71%	0.66	0.00%	2.08%
30–34	27	24	11	13	1	76
	36.99%	21.24%	9.41%	8.61%	3.70%	15.80%
35–39	21	35	35	52	6	149
	28.77%	30.97%	29.91%	34.44%	22.22%	30.98%
40–44	12	28	35	46	12	133
	16.44%	24.78%	29.91%	30.46%	44.44%	27.65%
45–49	4	17	21	27	7	76
	5.48%	15.04%	17.95%	17.88%	25.93%	15.80%
50–60	5	6	13	12	1	37
	6.85%	5.31%	11.11%	7.95%	3.70%	7.69%
Total	73	113	117	151	27	481
	100%	100%	100%	100%	100%	100%

Table A8. Percentage of Time Devoted to Research by Number of Years Spent Studying Abroad

Years Spent Abroad	0–30%		31–60%		61–99%		Total	
0	23	30.14%	33	45.20%	18	24.66%	74	100%
1–2	28	24.78%	57	50.44%	28	24.78%	113	100%
3–4	48	41.03%	53	45.30%	16	13.68%	117	100%
5–9	65	43.05%	64	42.38%	22	14.56%	151	100%
10–20	14	51.85%	10	37.04%	3	11.11%	27	100%
Total	178	36.80%	217	45.11%	87	18.09%	482	100%

Table A9. Frequency of Communication with International Research Assistance Organizations (Other than IFS) by Number of Years Spent Studying Abroad

Frequency of Communication	0 Years	1–2 Years	3–4 Years	5–9 Years	10–20 Years	Total
Never	34	42	47	54	7	184
	50.00%	39.62%	41.59%	39.13%	28.00%	40.89%
Rarely	20	35	34	42	7	138
	29.41%	33.10%	30.09%	30.43%	28.00%	30.67%
Annually	12	23	24	38	7	104
	17.65%	21.70%	21.24%	27.54%	28.00%	23.11%
Monthly	2	6	8	4	4	24
	2.94%	5.66%	7.08%	2.90%	16.00%	5.33%
Total	68	106	113	138	25	450
	100%	100%	100%	100%	100%	100%

Table A10. Tendency of Scientists to Work Alone or with Other Scientists by Number of Years Spent Abroad

No. of Years Spent Abroad	With Other Scientists		Alone		Total	
0	48	65.75%	25	34.25%	73	100%
1–2	95	84.07%	18	15.93%	113	100%
3–4	93	80.17%	23	19.83%	116	100%
5–9	111	73.03%	41	26.97%	152	100%
10–20	20	74.07%	7	25.93%	27	100%
Total	367	76.30%	114	23.70%	481	100%

Table A11. Maintenance of Professional Correspondence with
Foreign Scientists outside of International Scientific Meetings by
Number of Years Spent Abroad

No. of Years Spent Abroad	Yes		No		Total	
0	55	74.32%	19	25.68%	74	100%
1–2	91	81.98%	20	18.02%	111	100%
3–4	105	88.98%	13	11.02%	118	100%
5–9	136	90.07%	15	9.93%	151	100%
10–20	23	88.46%	3	11.54%	26	100%
Total	410	85.42%	70	14.58%	480	100%

Table A12. Frequency of Sabbatical Leave by Number of Years Spent
Studying Abroad

No. of Years Spent Abroad	Sabbatical Leave		No Sabbatical Leave		Total	
0	6	8.22%	67	91.78%	73	100%
1–2	22	19.30%	92	80.70%	114	100%
3–4	41	34.75%	77	65.25%	118	100%
5–9	38	25.00%	114	75.00%	152	100%
10–20	5	18.52%	22	81.48%	27	100%
Total	112	23.14%	372	76.86%	484	100%

Table A13. Frequency of Job Offers Abroad by Number of Years
Spent Studying Abroad

No. of Years Spent Abroad	Job Offers Abroad		No Job Offers Abroad		Total	
0	6	8.22%	67	91.78%	73	100%
1–2	15	13.27%	98	86.73%	113	100%
3–4	37	31.36%	81	68.64%	118	100%
5–9	54	35.76%	97	64.24%	151	100%
10–20	13	48.15%	14	51.85%	27	100%
Total	125	25.93%	357	74.07%	482	100%

Table A14. Frequency of Accepting a Job Offer by Number of Years
Spent Studying Abroad

No. of Years Spent Abroad	Offer Accepted		Offer Refused		Total	
0	0	0	6	100.00%	6	100%
1–2	3	20.00%	12	80.00%	15	100%
3–4	9	24.32%	28	75.68%	37	100%
5–9	17	38.48%	37	68.52%	54	100%
10–20	5	38.46%	8	61.54%	13	100%
Total	34	27.20%	91	72.80%	125	100%

References

Aguilar, J., and A. Vicente. 1987. *Papel del Estado en el desarollo de las disciplinas cientificas: el caso de Costa Rica.* San José: CONICIT.

Alvarez, R.D. 1984. *Universidad: Investigacion y productividad.* Caracas: Universidad Central de Venezuela.

Arunachalam, S. 1979a. "Why Is Indian Science Mediocre?" *Science Today* (Feb.): 8–9.

———. 1979b. "Scientific Journals in India: Their Relevance to International Science." *Science Today* (Mar. 1979): 45–50.

———. 1988. "The Links betweeen Mainstream Science and Journals on the Periphery." *Journal of Scientific and Industrial Research* 47 (June): 307–14.

Arunachalam, S. and K.C. Garg. 1985. "A Small Country in a World of Big Science: A Preliminary Bibliometric Study of Science in Singapore." *Scientometrics* 8, no. 5–6:301–13.

Arunachalam, S., and K. Manorama. 1988. "The Status of Scientific Journals of India as Seen through Science Citation Index." *Journal of Scientific and Industrial Research* 47 (July): 359–67.

Arunachalam, S., and S. Markanday. 1981. "Science in the Middle-Level Countries: A Bibliometric Analysis of Scientific Journals of Australia, Canada, India, and Israel." *Journal of Information Science* 3:13–26.

Arvanitis, R., and Y. Chatelin. 1988. "National Scientific Strategies in Tropical Soil Sciences." *Social Studies of Science* 18, no. 1:113–46.

AUPELF. 1984. *Répertoire des enseignants et chercheurs Africains, universités d'Afrique membres de l'AUPELF.* Montreal: Bibliotheque Nationale de Québec.

Baark, E. 1986. *The Context of National Information Systems in Developing Countries: India and China in a Comparative Perspective.* Lund, Sweden: Research Policy Institute, University of Lund.

Bailleuil, A. 1984. "L'Université de Dakar: Institutions et fonctionnement." Thèse de Doctorat d'Etat en Droit, Université de Dakar, 1984.

Barel, Y., and P. Malein. 1973. "Y a-t-il une profession de chercheur?" *La Recherche,* no. 39 (Nov.): 933–38.

Barrere, 1986. "La science en Inde." *La Recherche,* no. 180 (Sep.): 1136–48.

Basalla, G. 1967. "The Spread of Western Science." *Science* 156:611–22.

Beaver, D. de B., and R. Rosen. "Studies in Scientific Collaboration." Part 1, "The Professional Origins of Scientific Coauthorship." *Scientometrics* 1, no. 1 (1978): 65–84; Part II, "Scientific Coauthorship, Research Productivity and Visibility in the French Scientific Elite." *Scientometrics* 1, no. 2, (1979a): 133–49. Part III, "Professionalization and Natural History of Modern Scientific Coauthorship." *Scientometrics* 1, no. 3 (1979b): 231–45.

Berlinguet, L. 1980. *Report on an Exploratory Study on Practical Applications of IFS Projects.* Ottawa, Canada.

Bhagavan, M.R. 1984. *Technological Transformation of Developing Countries.* Stockholm, Sweden: SAREC Report No. 4.

Black, R.P., et al. 1972. "Data on Scientific and Technological Potential of Thailand," Vol. II. In *Report on Activities of Task on Scientific and Technological Potential of Thailand.* Stanford, Cal.: Stanford Research Institute.

Blickenstaff, J. and M.J. Moravcsik. 1982. "Scientific Output in the Third World." *Scientometrics* 4, no. 2: 135–69.

Blutstein, H.I., et al. 1981. *Costa Rica: A Country Study.* Washington, D.C.: American University, Foreign Area Study.

Bonnefond, P. 1987. *Notes Sénégalaises.* Document de travail No. 1. Paris: Département SDU, ORSTOM.

Bonnefond, P., and Ph. Couty. 1988. "Sénégal: Passé et avenir d'une crise agricole." *Tiers Monde* 29, no. 114 (Apr.-June): 319–40.

Botelho, A.J. 1983. *Les Scientifiques et le pouvoir au Brésil: Le Cas de la Société Brésilienne pour le Progrès de la Science (SBPS), 1948–1980.* Mémoire de DEA. Paris: STS/CNAM.

Bourdieu, P., and J.C. Passeron. 1964. *Les Héritiers.* Paris: Les Editions de Minuit.

Braibant, M. 1986. "Politiques macro-économiques et performances agricoles au Sénégal, 1960–1984." Thèse de Doctorat de 3ème cycle en économie du développement, Université de Paris I.

Braun, T., W. Glänzel, and A. Schubert. 1988. "The Newest Version of the Facts and Figures on Publication Output and Relative Citation Impact of 100 Countries, 1981–1985." *Scientometrics* 13, no. 5–6:181–88.

Bruer, J.T. 1984. "Women in Science: Toward Equitable Participation." *Science Technology and Human Values* 3–7 (Summer).

Busch, L. 1988. *Universities for Development: Report of the Joint Indo-U.S. Impact Evaluation of the Indian Agricultural Universities.* Lexington: Univ. of Kentucky.

Busch, L., and W.B. Lacy. 1983. *Science, Agriculture, and the Politics of Research.* Boulder, Colo.: Westview.

Cagnin, M.A., 1985. "Patterns of Research in Chemistry in Brazil." *Interciencia* 10:64–77.

Centre National de la Planification de la Recherche Scientifique. 1973. *Analyse du potential scientifique et technique au Sénégal.* Dakar: CNPRS.

Chatelin, Y., and R. Arvanitis. 1985. *Pratiques et politiques scientifiques.* Actes du Forum des 6 et 7 Février 1984. Paris: ORSTOM.

———— and ————. 1988a. *Stratégies scientifiques et développement: Sols et agricultures des régions chaudes.* Paris: ORSTOM.

———— and ————. 1988b. "National Scientific Strategies in Tropical Soil Sciences." *Social Studies of Science* 18: 113–46.

Chulalongkorn University. 1986. *Research in Chulalongkorn University.* Bangkok: Office of Research Affairs.

Clozel. 1916. *Circulaire au sujet de la création d'un Comité d'Etudes Historiques et Scientifiques de l'AOF.* Dakar, Ann. Mém. Et. Hist. et Sc. AOF.

Cole, J.R., and H. Zuckerman. 1983. "Marriage, Family, and Scientific Publication: Truth and Illusion in Science." Paper presented at the Macy Foundation Research Symposium on Women in Science, New York, Nov. 17–19.

Committee on International Migration of Talent (CIMT). 1970. *International Migration of High-Level Manpower.* New York: Praeger.

CONICIT. 1983. *Primer seminario sobre la situacion cientifica y tecnologica en Costa Rica.* San José.

————. 1984. *Diagnostico de la investigacion en Costa Rica.*

CONICIT/IDRC. 1982. *Situacion actual y caracteristicas de las actividades de investigacion en Costa Rica.* San Jose.

Cornevin, R., and Cornevin, M. 1964. *Histoire de l'Afrique.* Paris: Petite Bibliothèque.

Costa Rica: A Country Case Study. 1981. Washington, D.C.: GPO.

Costa Rica. 1985. "Dianostico del financiamiento externo de la Universidad de Costa Rica." Preparado por la Unidad de Financiamiento Externo, Vice Rectoria de Investigacion, Universidad de Costa Rica. Typed.

Costa Rica. 1986. *Programa nacional de ciencia y tecnologia.* San José: Ministry of Science and Technology.

Crane, D. 1972. *Invisible Colleges.* Chicago: Univ. of Chicago Press.

CUB. 1988. *Chulalongkorn University Bulletin.* Bangkok.

Davis, C.H. 1983. "Institutional Sectors of "Mainstream" Science Production in Sub-Saharan Africa, 1970–1979: A Quantitative Analysis." *Scientometrics* 5, no. 3:163–75.

Davis, C.H., and T.O. Eisemon. 1989. "Mainstream and Non Mainstream Scientific Literature in Four Peripheral Asian Scientific Communities." *Scientometrics* 15, no. 3–4:215–39.

Dedijer, S. 1958. "Windowshopping for a Research Policy." *Nature* (June 7): 367–71.

————. 1963. "Underdeveloped Science in Underdeveloped Countries." *Minerva* 2, no. 1:61–81.

Eisemon, T.O. 1979. "The Implantation of Science in Nigeria and Kenya." *Minerva* 12, no. 4:504–26.

————. 1982. *The Science Profession in the Third World.* New York: Praeger.

Eisemon, T.O., et al. 1982. "Transplantation of Science to Anglophone and Francophone Africa." *Science and Public Policy* 2, no. 4.

Eisemon, T.O., and C.H. Davis. 1989. "Publication Strategies of Scientists in Four Peripheral Asian Scientific Communities: Some Issues in the Measurement and Interpretation of Non-mainstream Science." In P.G. Altbach, et al., *Scientific Development and Higher Education: The Case of Newly Industrializing Nations.* New York: Praeger.

Elliot, H. 1984. *The Use of Administrative Data for Policy Analysis: Lessons from the Thai Department of Agriculture.* The Hague, Netherlands: ISNAR.

Encyclopaedia Universalis. 1988. "Les Chiffres du Monde."

Eres, B.K. 1982. "Socioeconomic Conditions Relating to the Level of Information Activity in Less Developed Countries." Ph.D. diss., Drexel University, Israel.

Eriksen, J.H., et al. 1988. *Kasetsart University in Thailand: An Analysis of Institutional Evolution and Development Impact.* A.I.D. Project Impact Evaluation Report. Washington, D.C.

Fondeville (de), A. 1986. "Communication sur le financement de la recherche." Seminar on the management of Senegalese research systems, June 23–27, Dakar.

Frame, J.D. 1985. "Problems in the Use of Literature-Based S&T Indicators in Developing Countries." In *Science and Technology Indicators for Development,* edited by H. Morita-Lou, 117–22. Boulder, Colo.: Westview.

Frame, J.D., F. Narin, and M.P. Carpenter, 1977. "The Distribution of World Science." *Social Studies of Science* 7:501–16.

Fuenzalida, E. 1971. *Investigación científica y estratificación internacional.* Santiago: Editorial Andrés Bello.

Gaillard, A.M. 1983. *Couples suédois vers un autre idéal sexuel.* Paris: Editions Universitaires.

Gaillard, J. 1979. "La Fondation Internationale pour la Science: Naissance d'une organisation de soutien aux jeunes chercheurs des PED." *Le Progrès Scientifique* no. 199–200 (Mar.-June): 55–68.

———. 1982. "Rapport de mission au Maroc pour la Foundation Internationale pour la Science." Stockholm, Sweden.

———. 1984a. *Réflexion sur une expérience concrète: La Fondation Internationale pour le Science.* Proceedings of "Forum Pratiques et Politiques Scientifiques," Feb. 6–7. Paris: Editions de l'ORSTOM, 139–42.

———. 1984b. "Rapport de mission à Madagascar pour la Foundation Internationale pour la Science." Stockholm, Sweden.

———. 1984c. "Les Productions Végétales." Paper presented at the Fourth IFS General Assembly, Rabat, Morocco, Oct.

———. 1985a. "L'Aide à la recherche aux jeunes chercheurs des PED et le rôle de la Fondation Internationale pour la Science." *Mondes en Développement* 13, no. 49:91–104.

———. 1985b. "Rapport de mission au Sénégal pour la Foundation Internationale pour la Science." Stockholm, Sweden.

———. 1986. "L'Aide étrangère et le financement de la recherche dans les pays en développement, *Liaison Bulletin,* no. 4, Dept H. June, 7–23. Paris: ORSTOM.

———. 1987. "Les Chercheurs des pays en développement." *La Recherche,* no. 189 (June): 860–70.

———, ed. 1988. *Politique, programmation, gestion de la recherche pour le développement.* Paris: IIAP.

Gaillard, J. and S. Ouattar. 1988. "Purchase, Use and Maintenance of Scientific Equipment in Developing Countries." *Interciencia* 13, no. 2:65–70.

Gaillard, J., and R. Waast. 1988. "La Recherche scientifique en Afrique." *Afrique Contemporaine, La Documentation Francaise,* no. 148:3–29.

————. 1989. "Les Chercheurs et l'émergence de communautés scientifiques nationales dans les pays en développement." Thèse de doctorat, CNAM, Paris.

Galina, C.S., and J.M. Russel. 1987. "Research and Publishing Trends in Cattle Production in the Tropics: Part 1. A Global Analysis." *CAB International 1987* 55, no. 10 (Oct.): 743–49.

Garfield, E. 1977, 1978, 1983. *Current Contents*, no. 17 (Apr. 11, 1977); no. 18 (May 29, 1978); no. 23 (Aug. 15, 22, 1983).

————. 1983. "Mapping Science in the Third World. *Science and Public Policy* 10, no. 3 (June): 112–27.

Gillet, J.E. 1976. *Analyse du potential scientifique et technique au Sénégal (1975/1976)*. Dakar: OGRST.

Glaser, W.A. 1978. *The Brain Drain, Emigration and Return*. Oxford: Pergamon Press.

Gleizes, M. 1985. *Un Regard sur l'ORSTOM 1943–1983*. Paris: ORSTOM.

Goma, L.K.H. 1968. "Some Obstacles to the Effective Utilization of Trained Scientific and Technical Personnel in Developing Countries." Paper presented at the eighteenth Pugwash Conference on Science and World Affairs, Nice, France.

Gomez, L.D., and J.M. Savage. 1983. "Searchers on That Rich Coast: Costa Rican Field Biology 1400–1980." In *Costa Rican Natural History*, edited by D.H. Janzen. Chicago: Univ. of Chicago Press.

Gomez, M.B., and V.V. Bermudez. 1979. *Encuesta sobre cientificos activos en Costa Rica (1978): Descripcion de la metodologia y presentacion de algunos resultados*. San Jose: CONICIT.

Gonzalez, L.F. 1976. *Historia de la influencia extranjera en el desenvolvimiento educacional y cientifico de Costa Rica*. San José: Editorial Costa Rica.

Gonzalez, O.F., and O.M. Calderon Alfaro. 1986. *Perfil occupacional de los profesores exbecarios de la Universidad de Costa Rica*. Sistema de Estudios de Postgrado. San Jose: UCR.

Gonzalez de Wong, M.I., and G. Macaya Trejos. 1986. *Diagnostico del regimen academico de la Universidad de Costa Rica."* San José: UCR.

Hargrove, T.R. 1979/80. "Communication among Asian Rice Breeders." *Journals of Research Communication Studies* 2:119–32.

Harmon, L.R. 1965. *Profiles of Ph.D.s in the Sciences*. Washington, D.C.: National Academy of Sciences.

Herzog, J.A. 1983. *Career Patterns of Scientists in Peripheral Communities*. Research Policy 12. 341–49. Amsterdam: Elsevier Science Publisher.

Honntrakul, L. 1953. *The Historical Records of the Siamese-Chinese Relations*. Bangkok: Mai Bithaya Press.

Idachaba, F.S. 1980. *Agricultural Research Policy in Nigeria*. Washington, D.C. IFPRI Research Report No. 17.

IDRC. 1982. "Resource Allocation to Agricultural Research." Proceedings of a workshop held in Singapore, June 8–10.

IFAN. 1961. "Historique de l'Institut Français d'Afrique Noire." *Notes Africaines*, no. 90 (Apr). Dakar.

ILO. 1974. *1974 World Population Year: Bulletin of Labor Statistics.* Geneva, Switzerland: International Labor Organization.

Ingram, J.C. 1971. *The Economic Change of Thailand, 1850–1970.* Stanford: Stanford Univ. Press.

ISNAR, 1981. *El sistema de investigación agropecuaria y transferencia de tecnologia en Costa Rica.* The Hague, Netherlands.

———. 1983. *La Recherche agronomique et zootechnique en Haute Volta.* Rapport d'une mission, Banque Mondiale/FAO/ISNAR/en Haute Volta, Mar.

———. 1988. *Organizational, Financial, and Human Resource Issues Facing West African Agricultural Research Systems.* Working Paper No. 9. The Hague, Netherlands.

ISRA, 1987. *Stratégies et programmation des recherches 1989–1993.* July. Dakar.

———. 1988. *Bulletin analytique documentaire—année 1986.* Dakar: ISRA.

ITCR. *Informe estadistico de los recursos humanos.* Cartago, Costa Rica: Departamento de Recursos Humanos, 1984.

Jiminiga, S. 1986. "Forces et faiblesses d'un systéme" *Afrique Nouvelle* (Oct.).

Kamchad, M. 1982. "The Beginning of a New Scientific Climate." *Bangkok Post,* Aug. 18.

Kamchorn, Manunapichu. 1981. "Professions in Pure Science Are Losing Popularity." *J. Sci. Soc. Thailand* 7 (1981):37–40.

Kidd, C.V. 1959. "Basic Research: Description versus Definition." *Science* 129: 368–71.

———. 1983. "The Movement of Younger Scientists Into and Out of the United States from 1967 to 1980: Some Aspects of the International Movement of Scientific Knowledge." *Minerva* 21, no. 4:387–409.

Kolinsky, M. 1985. "The Growth of Nigerian Universities, 1948–1980." *Minerva* 23, no. 1:29–61.

Krauskopf, M., and R. Pessot. 1983. *Actividad científica en Chile: Publicaciones registradas durante el período 1980–1982.* Archives de Biología y Medicina Experimentales no. 16, 17–27. Santiago, Chile.

Krauskopf, M., R. Pessot, and R. Vicuna. 1986. "Science in Latin America: How Much and Along What Lines?" *Scientometrics* 10, no. 3–4:199–206.

LaCoste, Y. 1959. *Les Pays sous développés.* Paris: PUF, Que sais-je? no. 853.

Lacy, W.B., L. Busch, and P. Marcotte. 1983. "The Sudan Agricultural Research Corporation: Organization, Practices, and Policy Recommendations." Lexington: Univ. of Kentucky. Mimeo.

Larue, B.M. 1988. "Quelques remarques concernant une expérience de coopération." Paris: ORSTOM. Typed.

Lawani, S.M. 1977. "Citation Analysis and the Quality of Scientific Productivity." *BioScience* 27, no. 1:26–31.

Legay, J.M. 1981. *Qui a peur de la science?* Paris: Editions Sociales.

Le Pair, C. 1986. *Some Comments on the Organization of Science and Technology in Thailand.* Utrecht, Netherlands: STW.

Lomnitz, L.A., M.W. Rees, and L. Cameo. 1987. "Publication and Referencing Patterns in a Mexican Institute." *Social Studies of Science* 17:115–33.

Maliyamkono, T.L., and S. Wells. 1980. "Effects of Training on Economic

Development: Impact Surveys on Overseas Training." In *Policy Developments in Overseas Training*, edited by T.L. Maliyamkono. Dar es Salaam, Tanzania: Black Star Agencies.

Martinez-Palomo, A., and H. Arechiga. 1979. "La Investigación biomédica en México. 1: La investigación básica." *Gaceta Médica de México*, no. 115:65–70.

Medawar, P.B. 1967. *The Art of the Soluble*. London: Methuen.

Mingsarn, S. 1981. *Technology Transfer*. Singapore University Press.

Monge, C. 1975. *La educación superior en Costa Rica*. San José, Costa Rica: Consejo Nacional de Rectores, Oficina de Planificacion de la Educacion Superior.

Moravcsik, M.J. 1976. *Science Development—The Building of Science in Less Developed Countries*, 2d ed. Bloomington, Ind.: PASITAM.

———. 1979. "Local Institutions for the Building of Science and Technology in Developing Countries." Paper prepared for a workshop organized by the American Association for the Advancement of Science, Washington, D.C., Apr.

———. 1985a. "Science in the Developing Countries: An Unexplored and Fruitful Area for Research in Science Studies." *Journal of the Society for Social Studies of Science* 3, no. 3:2–13.

———. 1985b. "Relevance: An Analysis Illustrated on Science Education in the Third World." *Interciencia* 10, no. 1:9–14.

———. ed. 1985c. *Strengthening the Coverage of Third World Science:* The Bibliographic Indicators of the Third World's Contribution to Science." Deliberations, Conclusions, and Initiatives of an Ad Hoc International Task Force for Assessing the Scientific Output of the Third World. Philadelphia, Penn., Jul.

Moravcsik, M.J., and S.G. Gibson. 1979. "The Dynamics of Scientific Manpower and Output." *Research Policy* 8:26–46. Amsterdam: Elsevier Science Publishers.

Morel, C.M., and R.L.M. Morel. 1978. "Estudo sobre a producão cientifica brasileira segundo os datos do Institute for Scientific Information (ISI): Um novo instrumento para a analise de ciência brasileira." *Ciência da Informação*, no. 7 79–83, Rio de Janeiro.

MOSTE. 1981. *Report on Scientific and Technical Manpower Survey*. Bangkok. In Thai.

———. 1983. *Scientific and Technical Manpower and R&D in the Private Sector*. Bangkok.

———. 1987. *The White Book on Science and Technology in Thailand*. Bangkok. In Thai.

MRST. 1981. "Enquête-inventaire du potentiel scientifique et technique au Sénégal." Survey conducted by the Ministry of Technical and Scientific Research.

MRST. 1984. *Le Centre National de Recherches Agronomiques de Bambey*. Bambey, Senegal.

Muscat, R.J. 1966. *Development Strategy in Thailand*. New York: Praeger.

National Science Foundation. 1981. *Science and Engineering Doctorates: 1960–1981*. NSF 83–309. Washington, D.C.: GPO.

———. 1984. *Women and Minorities in Science and Engineering*. Washington, D.C.

180   References

Nations Unies. 1980. *Programme d'action de Vienne pour la science et pour la technique au service du développement*. New York.

Ndiaye, G.H. 1986. "Les Composantes extérieures des politiques de recherche: Articulation du niveau national, régional, international: l'Exemple du Sénégal." Paper presented at the seminar on "Les Choix Stratéguiques d'une Politique de Recherche pour le Développement," IIAP, Paris, Sept. 22–26.

NESDB. 1981. *Framework for Future Science and Technology Development Plans*. Bangkok: Technology and Environment Planning Division.

Niang, S.M. 1987. *Africanisation totale du personnel dans trois ans."* Le Soleil, Dakar, Senegal, Jan. 4.

NRC. 1979. *Survey on University Graduate Manpower in Thailand*. Bangkok.

NRC/MOSTE. 1987. *Directory of Foreign Researchers' Research Projects in Thailand*. Bangkok.

ONEC. 1973. *Education Report, Institutions of Higher Education in Thailand*. Bangkok: Office of the National Education Commission.

Oram, P. 1985. "Donor Assistance to Agricultural Research: A Proposal for Information Exchange." Unpublished.

Oram, P., and V. Bindlish. 1981. *Resource Allocations to National Agricultural Research: Trends in the 1970s*. The Hague, Netherlands: ISNAR/IFPRI, Nov.

Packer, J.S., and W.P. Murdoch. 1974. "Publication of Scientific Papers in the Journals of the Entomological Society of America: an Eleven Year Review." *Bull. Entomol. Soc. Am.*, no. 20:249–53.

Pairor, Thipayasana. 1982. *The Development of Science in Thailand*. In Thai.

Paye, L. 1959. *Livret de l'Etudiant 1959–60*. Senegal: Université de Dakar.

Pernetta, J.C., and L. Hill. 1984. "The Role of Indigenous Professional and Scientific Organisations in the Development of the Pacific Region." In *Proceedings of 1984 Waigani Seminar*, University of Papua New Guinea.

Perrin, J. 1983. *Les transferts de technologie*. Paris: Editions la Découverte.

Porges, L. *Sources d'information sur l'Afrique Noire Francophone et Madagascar*. Paris: La Documentation Française.

Price, D.J. de Solla. 1963. *Little Science, Big Science*. New York: Columbia University Press.

Price, D.J. de Solla, and D. de B. Beaver. 1966. "Collaboration in an Invisible College." *American Psychologist* 21:1011–18.

Price, D.J. de Solla, and S. Gursay. 1975. "Some Statistical Results for the Numbers of Authors in the States of the United States and the Nations of the World." Preface to *Who Is Publishing in Science?* Philadelphia: Institute of Scientific Information.

Rabkin, Y., and H. Inhaber. 1979. "Science on the Periphery: A Citation Study of Three Less Developed Countries." *Scientometrics:* 261–74.

Raj, K. 1986. "Hermeneutics and Cross-Cultural Communication in Science: The Reception of Western Scientific Ideas in 19th Century India." *CNRS Revue de Synthèse*, Paris, 4, no. 1–2 (Jan.-June):107–20.

Ramirez, M.A. 1987. *Requerimientos de estudios de postgrado en Costa Rica, CONOCIT*. San José.

Report by the Committee of Vice-Chancellors of Australian Universities on the Situation in the University of the Pacific South (UPS). 1983. Canberra.

Richard, J. 1988. "Compte rendu de mission au Sénégal du 13 au 17 Mars 1988." Paris: Ministère de la Coopération. Typed.

Roche, M. 1966. "Aspects sociaux du progrès scientifique dans un pays en voie de développement." *IMPACT, Science et Société* 16, no. 1:53–63.

————. 1976. "Early History of Science in Spanish America." *Science* 194:806–10.

————. 1983. Interview with E. Mayz Vallenilla. "¿Abolir la investigación en la universidad?" *Interciencia* 8.

————. 1986. "¿Ha contribuido la ciencia al desarollo?" *Interciencia* 11, no. 5: 216–20.

————. 1987. "The Establishment of a Scientific Community in Developing Countries." Paper presented at the Second General Assembly of the Third World Academy of Science, Beijing, Sept. 14–18.

Roche, M., and Y. Freites. 1982. "Producción y flujo de información científica en un país periférico Americano (Venezuela) "*Interciencia* 7, no. 5:279–90.

Rodriguez-Sala de Gomez Gil, M.L., et al. 1970. *El científico en México.* Cuadernos de Investigación Social, No. 2. Mexico City: Free University of Mexico.

Rossi, G. 1973. "La science des pauvres" *La Recherche,* no. 30:7–15.

Rouille D'Orfeuil, H. 1987. *La Tiers Monde.* Paris: Editions La Découverte.

Ruellan, A. 1988. "La recherche scientifique: facteur de développement." *Le Monde Diplomatique,* Aug., 24.

Russel, J.M., et al. 1987. "Research and Publication Trends of a Latin American University." *Interciencia* 12:243–44.

————. 1981. "La universidad y el desarollo de la ciencia y tecnología. In *Ciencia, Technología y Desarollo Latinoamericano,* edited by F. Sagasti, 195–208. México: El Trimestre Económico/Fondo de Cultura Económica.

Sagasti, F., G. Oldham, G.V. Pisit, and P. Thiongane. 1983. *Evaluation of the International Foundation for Science* (1974–1981): Final Report. Stockholm: IFS.

Salam, A. 1966. "The Isolation of the Scientist in Developing Countries." *Minerva* 4, no. 4:461–65.

Salomon, J.J. 1970. *Science et politique.* Paris: Editions du Seuil.

————. 1984. "La science ne garantit pas le développement." *Futuribles* (June): 37–68.

————. 1986. "Science technologie et développement: Le Problème des priorités." Tiers Monde 27, no. 105 (Jan.-Mar.): 213–22.

Salomon, J.J., and A. Lebeau. 1988. *L'Ecrivain public et l'ordinateur: Mirages du développement.* Paris: Hachette.

Sanga, S., and Y. Yongyuth. 1984. "The Status and Quantitative Policy Targets of Science and Technology in Thailand." *ASEAN Journal on Science and Technology for Development* 1, no. 1:114–24. Singapore.

Santelices, B. *Problemas para hacer investigación en Chile: Una visión personal.* Santiago: Revista Chilena de Historia Natural. In press.

Sasson, A. 1986. *Quelles biotechnologies pour le Tiers-Monde?* Paris: Biofutur/ UNESCO.

Schoijet, M. 1979. "The Condition of Mexican Science." *Minerva* 12, no. 3: 381–412.

Schwartzman, S. 1978. "Struggling to Be Born: The Scientific Community in Brazil." *Minerva* 16, no. 4:545–80.

———. 1986. "Coming Full Circle: A Reappraisal of University Research in Latin America." *Minerva* 24, 4 (Winter): 456—75.

Sène, D. 1985. *Etude de l'impact de la recherche agronomique sur le développement agricole au Sénégal.* Paris: CIRAD.

Shenhav, Y.A., and D.H. Kamens. "The 'Cost' of Institutional Isomorphism." To be published in *Education and the Modern World,* edited by D.H. Kamens. London: Falmer Press.

Shiva, V., and J. Bandyopadhyay. 1980. "The Large and Fragile Community of Scientists in India." *Minerva* 18, no. 4:575–94.

Silo, 1983. *Sénégal.* Cahier d'Information No. 4. Paris: SILO, May.

Sirilli, G. 1986. *The Researcher in Italy: A Profession in Search of Recognition,* Research Policy No. 15:329–37. Amsterdam: Elsenier Science Publishers.

Soberon, G., and R. Mendoza de Flores. 1985. *La investigación y la universidad.* Revista de la Universidad Autónoma de México.

SOU. 1973. *Forskning för utveckling* (Research for Development). Stockholm: Statens Offentliga Utredningar 1973:41.

———. 1983. *Om hälften vore kvinnor* (If Half Were Women). Stockholm, Sweden: Lieberförlag.

Stepan, N. 1976. *Beginnings of Brazilian Science.* New York: Science History Publications.

Surajit Sinka. 1970. *Indian Scientists: The Socio-Cultural and Organizational Context of Their Profession in Science Technology and Culture.* New Delhi, India: India International Center.

Sylla, Y. 1986. "La situation de la recherche au Sénégal et le statut du chercheur." Lecture given at the Teachers' Training School, Dakar, Senegal.

TDRI. 1986. *A Computer Model for Resource Data Base.* Second Progress Report. Bangkok, Thailand: TDRI.

Thailand. National Statistics Office, Office of the Prime Minister. *School and Teacher Census, 1967 and 1968.* Final Report, Bangkok.

Thailand. Office of the Prime Minister. 1984. *Thailand in the 80s.* Thailand: National Identity Office.

Thai Life. 1985. "King Mongkut, the Father of Thai Science." *Science and Technology in Transition* 3, no. 2.

Thong, Saw Pak. 1968. "Malaysia: Problems of Science and Technology in Malaysia." Conference of Commonwealth Scientists, Royal Society, London.

Todaro, M.P. 1977. *Economic Development in the Third World.* London: Longman.

Tripier, P. 1984. "Approches sociologuiques du marché du travail: Essai de sociologie de la sociologie *(du travail).* "Thèse de doctorat d'Etat, Université Paris VII.

"Tropical Animal Production for Rural Development." 1988. A proposal to establish a computerized research journal to promote communication among scientists and decision makers concerned with rural development in the Third World. Buga, Columbia: Instituto Mayor Campesino.

Turner, W.A. 1984. "Quelques questions à propos des études bibliométriques." In *Pratiques et Politiques Scientifiques*, edited by Y. Chatelin and R. Arvanitis. Paris: ORSTOM.

UCR. 1985a. *Recursos de infraestructura en la Universidad de Costa Rica.* San José.

————. 1985b. *La producción de la investigación en la Universidad de Costa Rica.* San José.

————. 1985c. *Las caracteristicas del personal docente en la Universidad de Costa Rica.* San José.

UNESCO. 1971a. *Science for Development.* Paris.

————. 1971b. *Scientists Abroad: A Study of the International Movement of Persons in Science and Technology.* Paris.

————. 1985a. *Statistical Yearbook.* Paris.

————. 1985b. *Science and Technology in Countries of Asia and the Pacific*, No. 52. Paris.

UNESCO/Ministry of Education/Science Society of Thailand. 1987. *Five Years of Thailand's Youth Science Week (1982–1986).*

UNESCO/PNUD. 1985. *Inventaire du potentiel scientifique et technologique de la communauté.*

United Nations Vienna Program of Action for Science and Technology for Development. 1980. New York.

Velho, L. 1985. "Science on the Periphery: A Study of the Agricultural Scientific Community in Brazilian Universities." Ph.D. Thesis, SPRU, Univ. of Sussex.

————. 1986. "The Meaning of Citation in the Context of a Scientifically Peripheral Country." *Scientometrics* 9, no. 1–2:71–89.

Vessuri, H. 1985. "The Search for a Scientific Community in Venezuela: From Isolation to Applied Research." *Minerva* 22, no. 2:196–235.

————. 1986. "The Universities, Scientific Research and the National Interest in Latin America." *Minerva* 24, no. 1:1–38.

Weiss, P. 1960. "Knowledge: A Growth Process." *Science* 131:1716–19.

World Bank: 1986. *Report on World Development.* Washington, D.C.

Yuthavong, Y. 1979. "Cultural Factors in the Application of Science and Technology for Development." *Journal of the Science Society of Thailand* 5:55.

————. 1986. "Bibliometric Indicators of Scientific Activities in Thailand." *Scientometrics* 9, no. 3–4:139–43.

Zahlan, A.B. 1970. "Science in the Arab Middle East." *Minerva* 8, no. 1:8–35.

Zuckerman, H. 1978. "Theory Choice and Problem Choice in Science." *Sociological Inquiry* 48, no. 3–4:65–95.

Index

About the Author

Jacques F. Gaillard is coordinator of the Science, Technology, and Development program at the French Institute of Scientific Research in Cooperation for Development (ORSTOM). He obtained the degree of Agricultural Engineer from the Ecole Supérieure d'Agriculture d'Angers, France, in 1973 and worked as scientific secretary of the International Foundation for Science (IFS) in Stockholm, Sweden, from 1975 to 1985. He received his Diplôme d'Etudes Approfondies (1986) and his doctorate in Science, Technology, and Society from the Conservatorie National des Arts et Métiers (1989), Paris.

His professional duties since 1975 have brought him to some 80 countries in Europe, Africa, Asia, Latin America, and the Pacific. His present research interests include the science profession and the emergence of national scientific communities in developing countries, the comparative analysis of research policies for development, and the evaluation of research.